Blueprint Reading for Plumbers

RESIDENTIAL AND COMMERCIAL

5th edition

Blueprint Reading
for Plumbers
RESIDENTIAL AND COMMERCIAL

5th edition
J. Russell Guest
Bartholomew D'Arcangelo
Benedict D'Arcangelo

Delmar Publishers Inc.®

NOTICE TO THE READER

Publisher does not warrant or guarantee any of the products described herein or perform any independent analysis in connection with any of the product information contained herein. Publisher does not assume, and expressly disclaims, any obligation to obtain and include information other than that provided to it by the manufacturer.

The reader is expressly warned to consider and adopt all safety precautions that might be indicated by the activities described herein and to avoid all potential hazards. By following the instructions contained herein, the reader willingly assumes all risks in connection with such instructions.

The publisher makes no representations or warranties of any kind, including but not limited to, the warranties of fitness for particular purpose or merchantability, nor are any such representations implied with respect to the material set forth herein, and the publisher takes no responsibility with respect to such material. The publisher shall not be liable for any special, consequential or exemplary damages resulting, in whole or in part, from the readers' use of, or reliance upon, this material.

Delmar Staff

Associate Editor: Cameron Anderson
Editing Manager: Gerry East
Project Editor: Mary Ormsbee
Design Coordinator: Susan Mathews
Publications Coordinators: Judith Block, Karen Seebald

For information address Delmar Publishers Inc.
2 Computer Drive West, Box 15-015
Albany, New York, 12212

Printed in the United States of America
Published simultaneously in Canada
by Nelson Canada,
a Division of International Thomson Limited

Library of Congress Cataloging-in-Publication Data
Guest, J. Russell.
   Blueprint reading for plumbers: residential and commercial / J. Russell Guest, Bartholomew D'Arcangelo, Benedict D'Arcangelo. —5th ed.
   p. cm.
D'Arcangelo's name appears first on the earlier edition.
Includes index.
ISBN 0-8273-3459-1 (pbk.). ISBN 0-8273-3460-5 (instructor's guide)
1. Plumbing drafting. 2. Blueprints. I. D'Arcangelo, Bartholomew.
II. D'Arcangelo, Benedict. III. Title.
TH6231.D3 1989
696'.1'022'3—dc19   88-25668
                                             CIP

Contents

SECTION 4 COMMERCIAL BUILDING BLUEPRINTS

APPENDIX

RESIDENTIAL AND COMMERCIAL BLUEPRINTS

Preface

BLUEPRINT READING FOR PLUMBERS provides instructional material for plumbing and pipe fitting students who must develop the ability to interpret trade blueprints and to plan the installation of the required plumbing. This is accompanied by reference to the Building Officials and Code Administrators (BOCA) plumbing code for pipe size and material as well as acceptable fittings and fixtures. This revision brings the study up to date in code, pipe and fitting materials, and in fixture designs.

Upon completion of this course, the trainee should be able to interpret correctly all types of trade drawings, to make orthographic or isometric sketches of plumbing installations, and to make a mechanical plan of piping for either residential or commercial construction. Assignments at the end of each unit give students the opportunity for extensive practice and provide a means for the instructor to check student progress.

The content is organized in units of instruction grouped under four major topics:

- *Section 1 — Piping Drawings* reviews the kinds of drawings which the plumber must be able to interpret. It includes problems which involve the use of actual rough-in sheets.

- *Section 2 — Isometric Pipe Drawings* presents the principles and applications of isometric sketching. A variety of floor plans and construction situations, each requiring a different pipe layout, are used to give the student the opportunity to solve problems using isometric sketches.

- *Section 3 — Interpreting Residential Blueprints* provides the opportunity to master the reading of trade blueprints. Piping layouts are planned according to the requirements of the blueprints. The blueprints of a typical residence are supplied for use with these units.

- *Section 4 — Commercial Building Blueprints* is concerned with the special problems in commercial plumbing installations. A full set of blueprints of an actual two-story commercial building are supplied.

ABOUT THE AUTHORS

The authors who developed this instructional material have drawn on a wide background of experience in plumbing instruction in both day school and apprentice programs.

J. Russell Guest is a qualified mechanical engineer with experience in the building industry, manufacturing industry, and teaching and school administration. He has served as a curriculum specialist with the New York State Education Department on a number of curriculum projects, including plumbing, developed in cooperation with the Buffalo Board of Education. In addition to teaching plumbing mathematics in day school, he has taught mathematics and science to plumbing apprentices for the Plumbers Local #36, Buffalo, N.Y.

The late Bartholomew D'Arcangelo was a journeyman plumber, plumbing shop instructor, and plumbing estimator. He was a member and former officer of Plumbers Local #36, Buffalo, N.Y., the United Association of journeymen and apprentices of the Plumbing and Pipefitting Industry of the United States and Canada. He developed the plumbing shop courses, set up plumbing instruction in England and Germany for American soldiers, and helped write the apprentice study guides on which much of the present apprentice instruction is based.

The late Benedict D'Arcangelo was a journeyman plumber, plumbing shop instructor, and plumbing estimator. He gained wide experience in home, commercial, and industrial plumbing, including two years as supervisor for a firm doing water main plumbing. He was a member of the above union Local #36. He served as an officer, as the coordinator for the apprentice school and a member of the apprentice school board, and wrote part of the instruction for apprentices.

The authors hope that their combined years of experience will benefit instructors and students alike. This text has been used in its previous four editions in thousands of classrooms across North America. The fifth edition should prove equally as useful, both to first-time adopters and to those who have used and enjoyed the text over the years.

ACKNOWLEDGMENTS

Appreciation is expressed to the following individuals and companies for their help in providing material for this text:

Photographs and Rough-In Sheets

American Radiator and Standard Sanitary Corporation
American Standard, Incorporated
The Halsey W. Taylor Company
Nibco
U.S. Steel Corporation
Zurn Industries, Incorporated

Reviewers

Mr. John Pratt, Texas State Technical Institute
Mr. Andy Housand, Pensacola Junior College
Mr. Glenn Roundtree, The Construction School

Code Excerpts

BOCA International

SECTION 1

Pipe Drawings

Unit 1 BLUEPRINT READING AND SKETCHING

OBJECTIVES

After completing this unit, the student will be able to:

- explain why the plumber must read blueprints.
- explain why the plumber must know how to sketch plumbing features.

BLUEPRINTS SHOW THE BUILDING'S DESIGN

The architect, engineer, and drafter convey their ideas for a building's design through various working drawings. The drawings are duplicated so this information can be passed on to all who take part in the construction of the building. Duplicated drawings were originally called *blueprints* because they had white lines on a blue background. Today, even though other copying processes produce dark lines on a white background, these working drawings are still sometimes called blueprints.

Working drawings contain valuable information the plumber needs to know about the desired construction, figure 1-1. Working drawings often show where to locate plumbing work, the materials to use, and how to

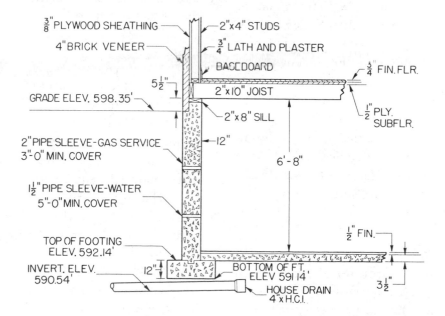

Fig. 1-1 Working drawings contain information about the desired construction.

install it. By reading the working drawings, the plumber learns how the designers want the job completed.

THE PLUMBER ON A CONSTRUCTION SITE

The plumber may be on the job from the start of construction. The storm sewer, sanitary sewer, and water service are often installed before other construction.

Plumbers have to accept responsibility for their work on the job. The number of plumbers on any one job may be few, perhaps only one plumber assisted by an apprentice. The plumber working independently is responsible for determining locations and elevations to insure a properly working plumbing system.

PLUMBERS USE SKETCHES

The design of a building does not include all of the details of the plumbing system. Some of the construction that the plumber and apprentice do is left to their planning. By consulting the working drawings and sketching the work to be done, plumbers can check on materials and locate difficulties on the project, figure 1-2. This saves time and labor.

Even when piping has been designed by others, it is helpful to make sketches of the parts. Generally, the full set of plans is kept in the field office because it is too cumbersome to carry to a specified point of installation. By doing a detailed sketch, the plumber has a ready reference to take to the work site. Sketches save plumbing time and help to avoid mistakes.

BUILDING CODES

The government control of construction is stated, or collected and printed, in booklets called *codes.* Building codes specify materials, sequence of construction, quality of work,

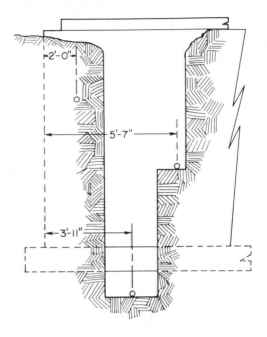

Fig. 1-2 Sketch of gas, water, and house drain locations from basement wall

and other details to insure a well-constructed and long-lasting building. *Plumbing codes* regulate details that keep the plumbing system running smoothly with minimum repair and maintenance. They cover such details as pipe sizes, materials, and venting.

The design and construction of a building has to follow the building codes if the inspector is to approve the final construction. The plumber is responsible for knowing the codes and keeping up with changes. Codes have changed over the years as new materials and connections are developed and as research changes construction methods.

ASSIGNMENT

Multiple Choice

Circle the correct answer for each question.

1. Where is the location of plumbing fixtures in a building shown?
   a. Plumber's sketches
   b. Working drawings
   c. Building codes
   d. Architect's plans

2. Who is responsible for locating pipes and fixtures correctly in a building?
   a. The apprentice
   b. The architect
   c. The master plumber
   d. The general contractor

3. When may plumbers start work on a building?
   a. Before any other construction begins
   b. After the roof is covered
   c. After the foundation is poured
   d. After the walls are erected

4. What is the purpose of building codes?
   a. To insure good quality
   b. To insure long-lasting plumbing systems
   c. To reduce maintenance
   d. All of the above

5. On what does the building inspector base the inspection of a structure?
   a. The workmanship
   b. The building codes
   c. The architect's plans
   d. The plumber's sketches

Unit 2 PIPE DRAWINGS

OBJECTIVES

After completing this unit, the student will be able to:

- describe the types of drawings used by the plumber.
- discuss the purposes of specifications and plumbing codes.

WORKING DRAWINGS

The working drawings convey information about the construction to the builder. These drawings try to give a three-dimensional view of the building. Orthographic and isometric drawings and mechanical plans make up the working drawings. The plumber must interpret the working drawings to install the plumbing system properly.

ARCHITECTURAL SYMBOLS

Architectural symbols represent various features in a construction. There are symbols to represent electrical, plumbing, heating, and ventilating features; to indicate the type of materials used; and to show structural features, such as windows and doors. The appendix has a list of architectural symbols commonly used on working drawings.

ORTHOGRAPHIC DRAWINGS

Orthographic drawings view a building by looking directly at the front, top, and side of it. These views are drawn to scale. All lines are true lengths and angles are not distorted.

Orthographic drawings are made of the elevations, floor plans, section views, and detail drawings of a building. *Elevations* show the front, rear, and side exterior views

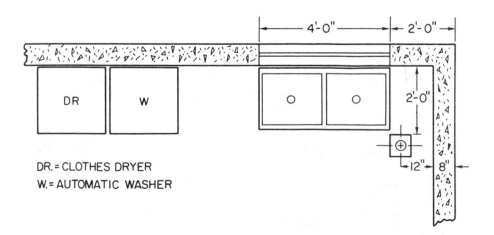

DR. = CLOTHES DRYER
W. = AUTOMATIC WASHER

Fig. 2-1 A portion of a basement floor plan is shown in a detail drawing.

of the building. By passing a horizontal cut midway between each floor, a *plan view* or *floor plan* is seen. This cut passes through all doors, windows, and wall openings to show the room as it appears when looking directly down on it. If a vertical cut is made through the building, then a section view is shown. *Section views* show the interior of the building from the foundation to the roof when looking directly at its side. Since all views are drawn to scale, certain building features may be too small to see its construction. In this case a *detail drawing* is made by enlarging the feature in scale, figure 2-1.

MECHANICAL PLANS

The location of piping in a building is sometimes shown on a plumbing *mechanical plan,* figure 2-2. The mechanical plan is made by drawing centerlines of pipe on floor plans of the building. On large buildings, the architect may supply these drawings or the plumber may make a mechanical plan to help estimate and assemble the job.

ISOMETRIC DRAWINGS

Isometric drawings combine the front, top, and side views of the orthographic projection into a single picture to give a three-dimensional effect. The vertical lines remain vertical in isometric. The horizontal lines are projected at 30-degree angles. The lengths of the lines remain the same as in orthographic drawings.

The plumber makes isometric pipe diagrams to help visualize what is required to install the piping, figure 2-3, page 6. The isometric drawing shows pipe sizes and the fittings required. It is possible to estimate

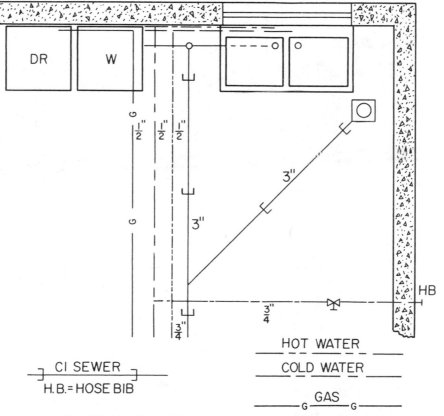

Fig. 2-2 Mechanical plan for a portion of a basement

from the isometric drawing as well as to convey information to another plumber. Plumbing permits and inspection are aided by the use of isometric drawings.

SPECIFICATIONS

The architect's drawings would be too complicated if all descriptive information were placed on them. That is why part of the description is written in the *specifications*. The specifications regulate the type or grade of material, the quality of the work, and the work to be done by the general contractor in relation to the work of the plumber. Typical specifications for a commerical building are shown in the appendix.

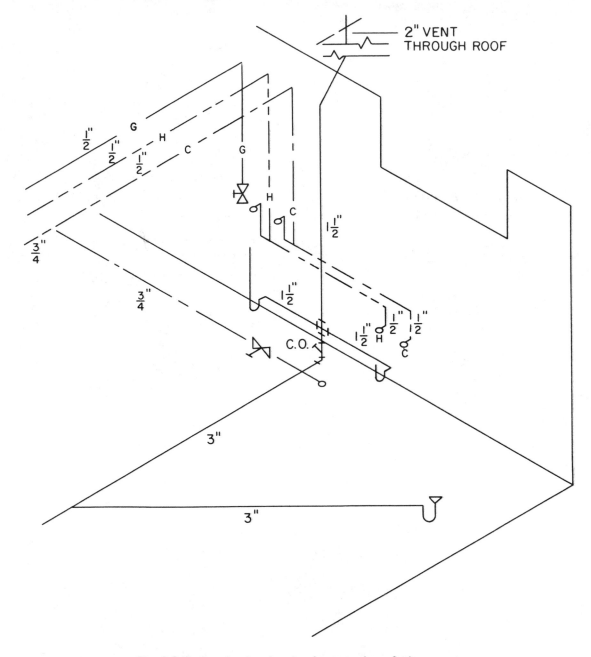

Fig. 2-3 Isometric pipe drawing for a portion of a basement

The specifications help the plumber to understand the architect's instructions. They modify the architect's plans and include information not readily seen on them. They may even change what is shown on the plans. Whatever is shown on the plans or called for in the specifications, whichever is greater, is part of the contract.

THE PLUMBING CODE

All plumbing work is controlled by a general set of regulations called the plumbing code. This code varies from one community to another in some details. The purposes are the same in every locality. The *plumbing code* is a set of written plumbing requirements to protect public health, insure long-lasting material, and require good workmanship.

The appendix contains paragraphs from the BOCA National Plumbing Code. It is compiled and published by Building Officials & Code Administrators International, Inc. (BOCA). The provisions of this code are in general use. The local codes may have additional provisions to reflect the needs of weather conditions and other experience in plumbing. The BOCA National Plumbing Code is used in this text where it applies to blueprint reading and sketching problems.

ASSIGNMENT

Multiple Choice

Circle the correct answer for each question.

1. Which two dimensions are indicated on a floor plan?
   a. Width and length
   b. Width and height
   c. Length and height
   d. Length and angle

2. A front elevation shows which two dimensions?
   a. Width and length
   b. Width and height
   c. Length and height
   d. Length and angle

3. A side elevation shows which two dimensions?
   a. Width and length
   b. Width and height
   c. Length and height
   d. Length and angle

4. In an isometric drawing, what is not correct in relation to the actual construction?
   a. Scale lengths
   b. Angles
   c. Width and length
   d. Width and height

5. What is not part of the specifications?
   a. Pipe material
   b. Workmanship required
   c. Responsibility of the general contractor
   d. Dimensions of the bathroom

6. What determines whether galvanized iron, copper, or plastic water pipe is to be used?
   a. Specifications
   b. Working drawings
   c. Isometric sketch
   d. Mechanical plan

7. Which plumbing code would apply to the construction of an office building?
   a. Federal code
   b. State code
   c. Local community code
   d. All of the above

8. What is not a purpose of the plumbing code?
   a. To assure correct piping installation
   b. To prevent the use of new products
   c. To protect public health
   d. To assure good workmanship

9. Which are the most informative drawings to file with the plumbing inspector?
   a. Mechanical plans
   b. Working drawings
   c. Isometric pipe drawings
   d. Plumbing sketches

10. How does a plumber *visualize* a drawing?
    a. By locating the plumbing fixtures
    b. By finding the overall dimensions
    c. By making a mechanical plan
    d. By forming a mental picture of the design

Unit 3 READING THE ARCHITECT'S SCALE

OBJECTIVES

After completing this unit the student will be able to:
- use the architect's scale.
- measure scale lengths.

SCALE DRAWINGS

Review scale drawings using the explanation contained in Unit 3 of *Basic Construction Blueprint Reading.* Be sure to note the explanations for each of these points:

- A scale drawing represents a large building on a rather small piece of paper.
- A scale drawing is measured correctly.

- The architect's scale simplifies the making of scale drawings.
- A 6-foot rule is used to measure a drawing on the job.

THE PLUMBER AND SCALE MEASUREMENT

There are two skills a plumber needs when using an architect's scale. The plumber

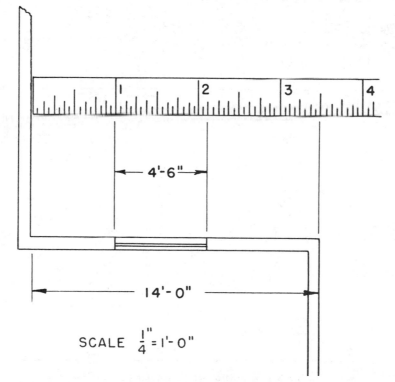

Fig. 3-1 Scale measure with a rule

4'-6"

14'-0"

SCALE $\frac{1}{4}" = 1'-0"$

must be able to (1) use an architect's scale to make a drawing, and (2) know how to measure a set of plans.

On the job, the plumber may use a regular rule for checking measurements on a drawing, figure 3-1. The regular rule is not as accurate as an architect's rule. However, the plumber must master using both on the job.

ASSIGNMENT

Drawing Exercise

Make a drawing of a portion of the basement plan, figure 3-2, to a scale of 1/2″ = 1′-0″. Figure 3-2 shows most details but is not drawn to scale.

Note: A poured concrete wall can be made almost any thickness for most house construction. A wood wall is usually framed with 2″ x 4″ studs. A 2″ x 4″ piece of lumber is 1 1/2″ x 3 1/2″ after being seasoned and dressed.

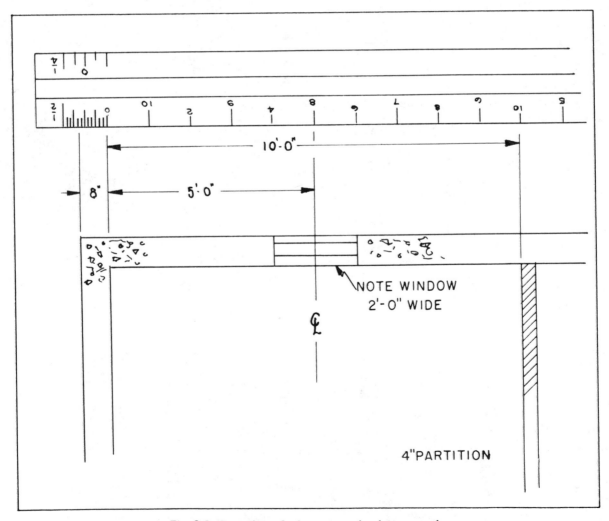

Fig. 3-2 A portion of a basement plan is measured.

Multiple Choice

The following questions are based on the scale drawing of figure 3-2. Circle the correct answer.

1. If the partitions in the basement are built with 3 1/2 inches of framing and covered with 3/8-inch hardboard panels on both sides, what is the full wall thickness?
   a. 4 3/4 inches
   b. 3 5/8 inches
   c. 4 1/4 inches
   d. 4 5/8 inches

2. How thick is a partition with the studs used flat (1 1/2-inch) and covered with 1/4-inch hardboard on both sides?
   a. 2 1/4 inches
   b. 1 3/4 inches
   c. 4 inches
   d. 2 inches

3. What is the greatest diameter fitting that could be concealed in the wall described in question #1?
   a. 3 1/2 inches
   b. 4 inches
   c. 3 7/8 inches
   d. 2 inches

4. What is the greatest diameter fitting to be concealed in the wall decribed in question #2?
   a. 2 inches
   b. 1 1/2 inches
   c. 3 1/2 inches
   d. 4 inches

5. How long will 8'-6" measure on an architect's rule if a scale of 1/2" = 1'-0" is used?
   a. 8 1/2 inches
   b. 4 1/4 inches
   c. 17 inches
   d. 16 1/2 inches

Scale Readings

Give the missing dimensions in figure 3-3 of the architect's scale.

A= _____   H= _____

B= _____   I= _____

C= _____   J= _____

D= _____   K= _____

E= _____   L= _____

F= _____   M= _____

G= _____

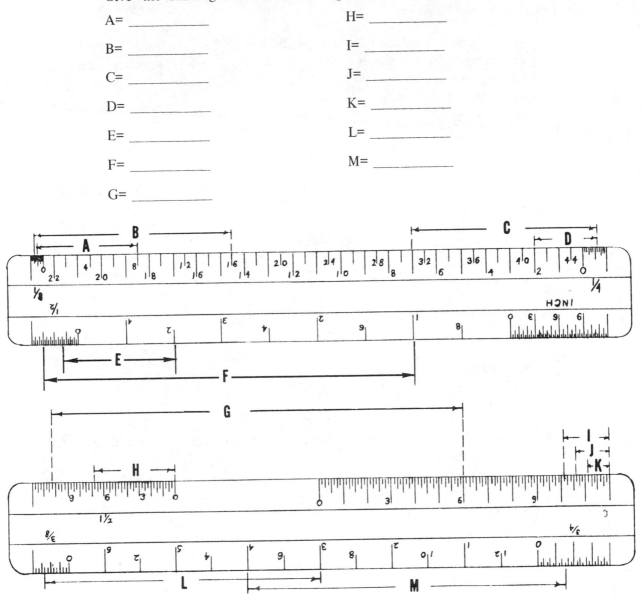

Fig. 3-3

Unit 4 LAUNDRY TRAYS AND FLOOR DRAINS IN BASEMENT PLANS

OBJECTIVES

After completing this unit, the student will be able to:

- discuss dimensions and materials of laundry trays and how they are shown in plan views.
- discuss materials, construction, and pipe connections for a floor drain.

LAUNDRY TRAYS

The standard laundry tray has a double compartment, figure 4-1. A common size is 2'-0" x 4'-0", figure 4-2.

The standard laundry tray is indicated on the floor plan unless there is a space problem or a special design is used. Where space is at a premium, or when an automatic washing machine is to be installed, the single basin laundry tray may be used. The automatic washer can also have plumbing without a laundry tray.

Laundry trays are made of fiberglass as well as precast concrete. They come in a variety of sizes. The floor plan shows the standard symbol for the laundry tray, figure 4-3, page 14. The actual description and details are included in the specifications. The laundry tray fixture is specified by the manufacturer's name and catalog number.

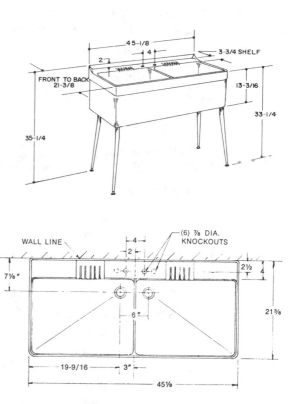

Fig. 4-1 Standard, double-compartment laundry tray

Fig. 4-2 Overall dimensions of a laundry tray

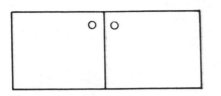

Fig. 4-3   Floor plan symbol for a double-compart-ment laundry tray

FLOOR DRAINS

The *floor drain* is a fitting set in the concrete floor to permit water to enter the house drain and flow to the sewer, figure 4-4.

A trap is installed between the house drain pipe and the floor drain to prevent sewer gases from entering the house. The trap used in residences is often a 3-inch, cast iron P trap.

The floor drain is also made of cast iron and has a 9-inch diameter or a 9-inch square top surface. Only the drain plate shows in the finished floor.

The manufacturers' catalogs show the details for variations of design. On the floor plan, the symbol for floor drains, except special purpose floor drains, is the circle in a square, figure 4-5.

A Pipe Size In	DIMENSIONS IN INCHES					Open Area Sq. In.
	B	C	E	F	K	
4,5,6	9	12	3	5 1/8	2	17
8	9	12	3	5 5/8	2	17

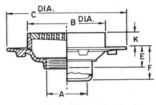

Fig. 4-4 Floor drain

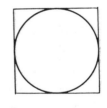

Fig. 4-5 Plan symbol for a floor drain

ASSIGNMENT

Drawing Exercise

Make a neat plan view of the laundry room in figure 4-6. Use a scale of 3/8″ = 1′-0″.

Multiple Choice

The following questions refer to the plan view in figure 4-6 and the elevation view in figure 1-1. Circle the correct answer for each question.

1. What is the purpose of the footing under the cellar wall?
    a. It is a firm, level surface on which to build the wall.
    b. The footing reduces settling.
    c. The footing is a support for the wall forms.
    d. All of the above
2. Where is the level of the finished cellar floor?
    a. Above the footing
    b. Even with the top of the footing
    c. At the base of the footing
    d. None of the above

3. Where is the house drain in relation to the footing?
   a. Above the footing   c. Below the footing
   b. Through the footing   d. None of the above

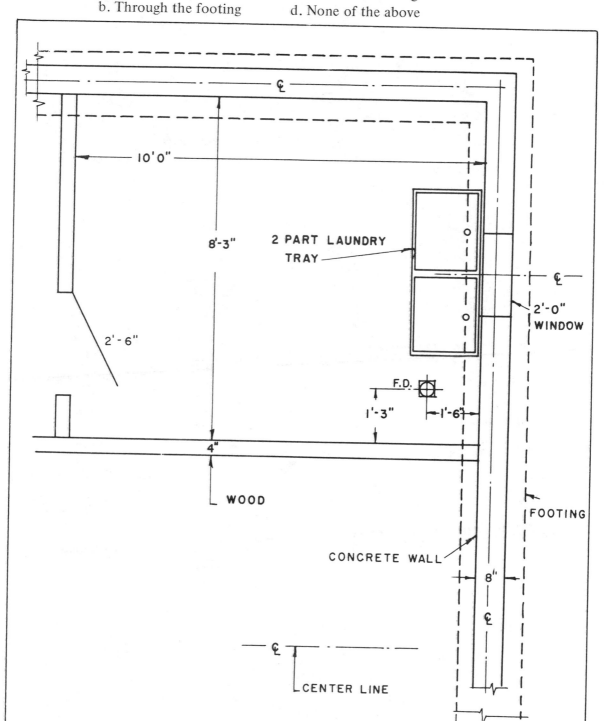

Fig. 4-6

Unit 5 KITCHEN FLOOR PLANS

OBJECTIVES

After completing this unit, the student will be able to:

- identify the floor plan symbols for sinks and other kitchen equipment.
- describe the details of kitchen planning.

KITCHEN DESIGN

The three principal units in a kitchen are the sink, range, and refrigerator. Each is represented on the floor plan by a rectangle with an identifying callout, figure 5-1.

In most floor plans the sink is located under a window. It is either built into a cabinet or set in a cabinet of its own. The built-in dishwasher is located near the sink. The base cabinets are generally 24 inches wide and have a moisture-proof working surface. The dish or wall cupboards are about 12 inches wide. Note how this is shown in the plan view.

Modern kitchen designs may use built-in cooking units and refrigerators. These may have somewhat different symbols from those used for the floor-standing range and refrigerator.

KITCHEN SINKS

Kitchen sinks may be either a single-compartment or double-compartment style, figures 5-2 and 5-3. Double-compartment sinks may have one deep compartment to combine a kitchen sink and a laundry tray. Sink fittings include a combination faucet and sometimes a dish spray on a hose. In each case a hot and cold water supply is needed.

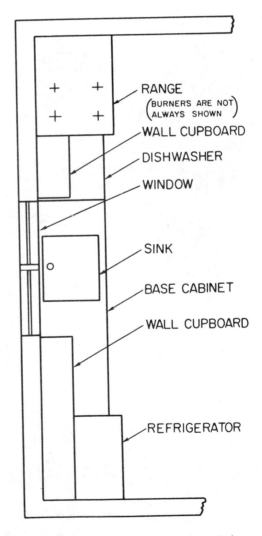

Fig. 5-1 Kitchen equipment symbols

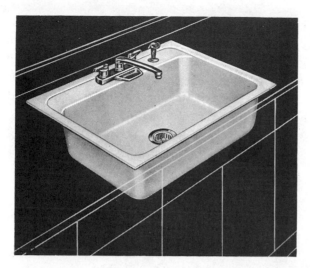

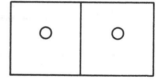

Fig. 5-2 Single-compartment sink and plan symbol

Fig. 5-3 Double-compartment sink and plan symbol

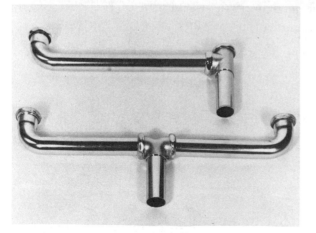

Fig. 5-4 Twin waste openings

Fig. 5-5 Sink strainer

Double-compartment sinks can be installed with one waste opening so that both basins discharge through the same trap. This is called a twin waste, figure 5-4. Two waste openings and two traps can also be used. The dishwasher is connected to the waste at a connection above the trap. Some sinks are equipped with a grinder (garbage disposal unit) in the waste opening to dispose of garbage to the sewer. However, the basket in each sink strainer can prevent garbage from entering the plumbing system, figure 5-5.

The manufacturers' catalogs have information on sink styles and sizes. A single-compartment sink can be 30″ x 21″. A double-compartment sink can be 42″ x 21″.

ASSIGNMENT

Blueprint Reading

Study the kitchen plan in the residential drawings contained in Packet 1 supplied with this text. Complete the following.

1. Sketch the symbol for the kitchen sink.

2. Are the cupboards over the refrigerator? _____

   Give reasons for your answer. _____

   _____

3. According to the residential plans, what floor size refrigerator can be installed in the space shown? _____

Drawing Exercise

Draw a neat plan of the kitchen shown in figure 5-6. Use a scale of 1/4″ = 1′-0″. This plan will be used for the Assignment in Unit 7.

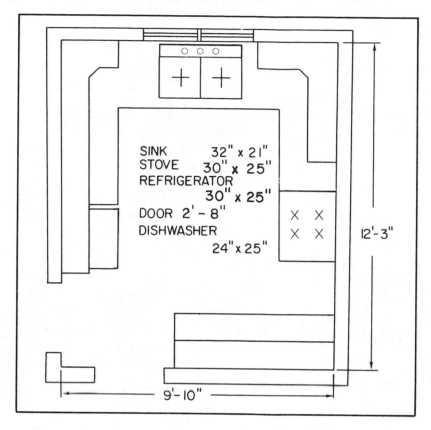

SINK   32" x 21"
STOVE   30" x 25"
REFRIGERATOR
              30" x 25"
DOOR 2' – 8"
DISHWASHER
              24" x 25"

12'-3"

9'-10"

Fig. 5-6

Unit 6 BATHROOM FLOOR PLANS

OBJECTIVES

After completing this unit, the student will be able to:

- identify the floor plan symbols for bathtub, water closet, lavatory, and shower.
- describe details of bathroom planning.

BATHROOM DESIGN

The usual bathroom has a door, a window, and three fixtures—a bathtub, water closet, and lavatory. It is not considered good design to have the tub under the window or where pipes will be in the outside wall, although this is often done to conserve floor space. Bathrooms are designed in many sizes. Where space is important, the minimum size is about 5'-0" x 7'-0".

BATHTUBS

Built-in bathtubs require a space of 5'-0" x 2'-6". There is some variation in width with design. Special tubs may be different sizes. This information is supplied by the architect or from the manufacturers' specifications.

Tubs have either right-hand or left-hand outlets depending on the location of the drain as one faces the installed tub. The outlet end of the tub should be located so that an access panel can be installed on the opposite side of the wall, figure 6-1. By removing the holding screws, it is possible to open the access panel for plumbing repairs.

The built-in tub with its supply and waste piping is installed while the wall framing is exposed. Therefore, the tub must be protected from damage during construction. One way is to paste paper on all the finished surfaces of the tub.

WATER CLOSETS

The water closet consists of a tank and a bowl. Both are made of vitreous china. They must be handled carefully as they can be broken by rough usage.

The floor plan in figure 6-2 shows a water closet. The tank is shown as a rectangle,

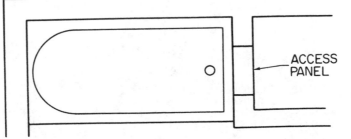

Fig. 6-1 Built-in bathtub

usually 20″ x 8″. Recent practice is to indicate the bowl as a rectangle rather than an oval or circle. The bowl rectangle is about 14″ x 20″. The distance from the wall to the front of the bowl is 2′-6″, although designs vary.

The soil stack is located in the wall near the toilet. In well-drawn plans, the wall is thicker in order to conceal the stack. Sometimes 2″ x 6″ studs are used instead of 2″ x 4″ studs to make it easier to accommodate the stack. Six-inch lumber is dressed to a 5 1/2-inch width.

LAVATORIES

The bathroom lavatory (wash basin) may be several different designs and sizes. It may be made of porcelain over steel or cast iron, vitreous china, or simulated marble. Many lavatories are now built into a cabinet or vanity. Others have legs and pedestals to give extra support and to vary the design.

A common size lavatory is 20 inches long and 18 inches from front to wall. In the floor plan in figure 6-3, the lavatory is shown as a rectangle.

SHOWERS

Many bathrooms are designed with a shower stall instead of a tub, figure 6-4. These showers are built on the job. Their walls are made of framing, plaster, and tile. Usually such a shower needs a floor space of 3′-0″ x 3′-0″. Showers can also be bought

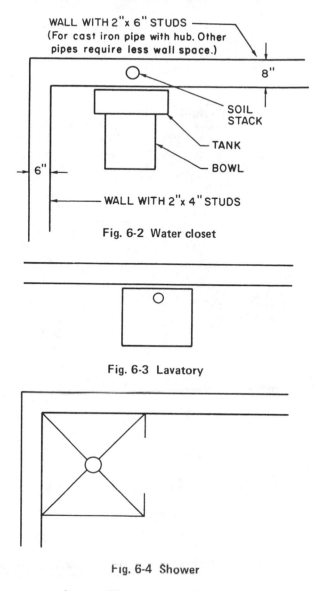

Fig. 6-2 Water closet

WALL WITH 2″ x 6″ STUDS
(For cast iron pipe with hub. Other pipes require less wall space.)

8″

SOIL STACK

TANK

BOWL

6″

WALL WITH 2″ x 4″ STUDS

Fig. 6-3 Lavatory

Fig. 6-4 Shower

as a package. The receptacle is set in place and the wall panels are assembled to form the shower enclosure. Such showers are often 2′-6″ x 2′-6″.

ASSIGNMENT

Blueprint Reading

Study the house floor plans supplied in Packet 1 for the bathroom layouts in the basement and first floor.

1. Make a list of the plumbing fixtures needed for both bathrooms.

2. What size soil (vent) stack is shown on the plans? _____

3. A dressed 2″ x 6″ stud gives how wide a space in the wall to enclose the vent stack? _____

Drawing Exercises

1. Make a neat floor plan layout for the bathroom sketch with a tub in figure 6-5. Use the proper symbols and a scale of 1/4″ = 1′-0″.

2. Make a neat floor plan layout for the bathroom sketch with a shower stall in figure 6-6. Use the proper symbols and a scale of 1/4″ = 1′-0″.

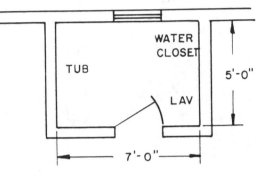

Fig. 6-5 Bathroom with tub

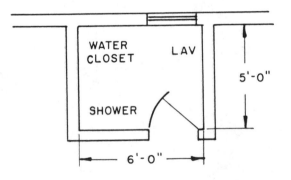

Fig. 6-6 Bathroom with shower stall

Unit 7 ROUGH-IN FOR KITCHEN SINK

OBJECTIVES

After completing this unit, the student will be able to:

- discuss the purpose of rough-in sheets.
- interpret the rough-in sheet correctly.
- use the rough-in sheet to make an elevation view of the water supply and waste rough-in locations.

ROUGH AND FINISH PLUMBING

The plumbing in a house consists of two general operations—rough plumbing and finish plumbing. In rough plumbing, the piping for soil stack, waste, vent, hot and cold water, and gas is installed. The water pipe openings are capped and the lines tested for leaks. The soil stack, waste, and vent lines are tested by using special plugs in the openings and filling the system with water. After inspection, the piping is hidden by lath and plastering or drywall on the walls.

The finish plumbing consists of installing the fixtures with the trim, such as faucets, etc. Built-in tubs are installed before the wall surface is finished, but the tub trim is installed during the finish plumbing.

ROUGH-IN SHEETS

Rough-in sheets show the plumber where to install waste and water supply pipes. These drawings also contain other information, such as how to install the fixture. Rough-in sheets are obtained from the manufacturer of the fixture and apply only to that fixture. Figure 7-1, page 24, is an example of a rough-in sheet for a kitchen sink.

ROUGH-IN FOR KITCHEN SINK

The rough-in sheet for the kitchen sink shows where the hot and cold water pipes emerge from the wall. These pipes should come out of the wall so that an elbow and a nipple point directly at and are below the end of the faucet connection shown in the rough-in sheet. Allow at least 6 inches below the faucet end to the center of the water pipe emerging from the wall.

The tailpieces of the double-compartment sink can each be connected to a trap and each trap to a waste connection emerging from the wall. It is also possible to connect both tailpieces by using a twin waste and a single trap with a waste opening.

For a double-compartment sink and two traps, the waste openings are about 22 inches above the finished floor. The same height is used for a single-compartment sink. Twin waste connections are made with either an end outlet or a center outlet. The twin waste opening is about 17 inches above the floor.

A garbage disposal installation is simplified where the individual waste traps and outlets are used with a double-compartment sink.

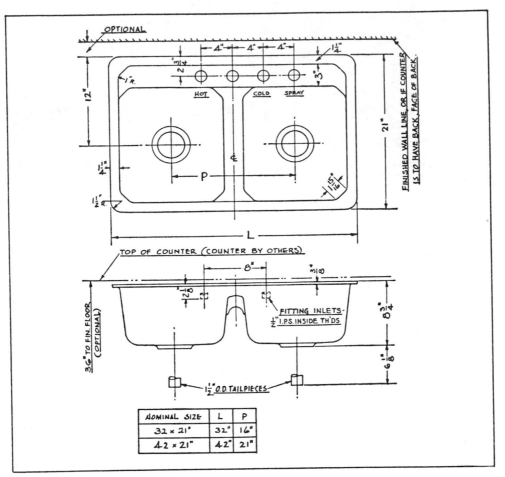

Fig. 7-1

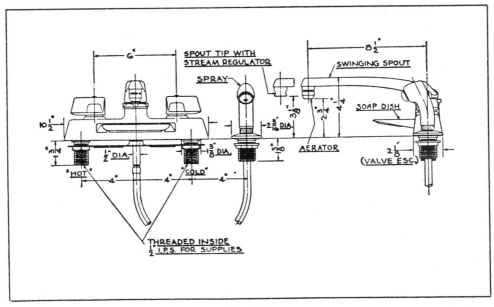

Fig. 7-2

ASSIGNMENT

Drawing Exercise

Use the kitchen plan drawn for the drawing exercise in Unit 5. Make an elevation of the wall against which the sink is located. Use a scale of 1″ = 1′-0″. Show the window and sink counter height. Locate the rough-in points for the waste outlet and hot and cold water. Use the sink described in figure 7-1 and 7-2.

Unit 8 ROUGH-IN FOR LAVATORY

OBJECTIVES

After completing this unit, the student will be able to:

- discuss the installation of a wall-hung lavatory.
- determine the locations of the hanger, backing, and waste and water opening.

HANGER

The wall-hung lavatory, like many other wall-hung fixtures, is supported by a hanger (see figure 8-1). This is a metal support fastened to the wall by screws. For this type fixture, the hanger is installed on the finished wall. There are lugs on the back of the lavatory that slip into the hanger.

BACKING

The backing consists of wood pieces either set into the studs or nailed between the studs to support the hanger and the lavatory basin. The backing is installed at the time the rough plumbing is done because it is behind the finished wall. It is the plumber's job to see that the backing is in place.

TUBING TRAP

Chrome-plated brass tubing trap is connected to the tail pipe of the lavatory waste by means of a packing nut shown on the rough-in sheet. This packing nut compresses a rubber ring or gasket to make a leak-tight connection.

The trap is connected to the waste pipe by methods approved in local codes. Either 1 1/4-inch or 1 1/2-inch traps can be connected to the 1 1/2-inch waste pipe fitting.

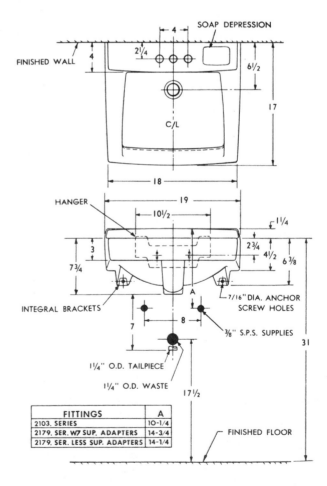

FITTINGS	A
2103. SERIES	10-1/4
2179. SER. W7 SUP. ADAPTERS	14-3/4
2179. SER. LESS SUP. ADAPTERS	14-1/4

Fig. 8-1 Vitreous china lavatory

FAUCETS

The faucets are installed on the lavatory by using putty or plastic element for a cushion and sealer. The hold-down nuts are tightened by the use of a basin wrench. The faucets are connected to the hot and cold water supply by flexible tubing and connectors.

ASSIGNMENT

Multiple Choice

The following questions are based on the rough-in sheets for the vitreous china lavatory, figure 8-1; the P trap and compression trap adapter, figure 8-2; and the combination lavatory fittings, figure 8-3. Circle the correct answer for each question.

1. How far above the finished floor is the rim of the lavatory?
   a. 32 1/4 inches　　c. 31 inches
   b. 28 1/4 inches　　d. 26 1/2 inches

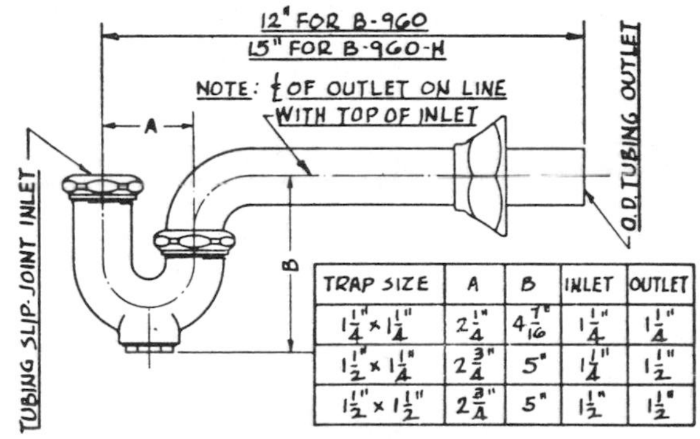

TRAP SIZE	A	B	INLET	OUTLET
1 1/4" x 1 1/4"	2 1/4"	4 7/16"	1 1/4"	1 1/4"
1 1/2" x 1 1/4"	2 3/4"	5"	1 1/4"	1 1/2"
1 1/2" x 1 1/2"	2 3/4"	5"	1 1/2"	1 1/2"

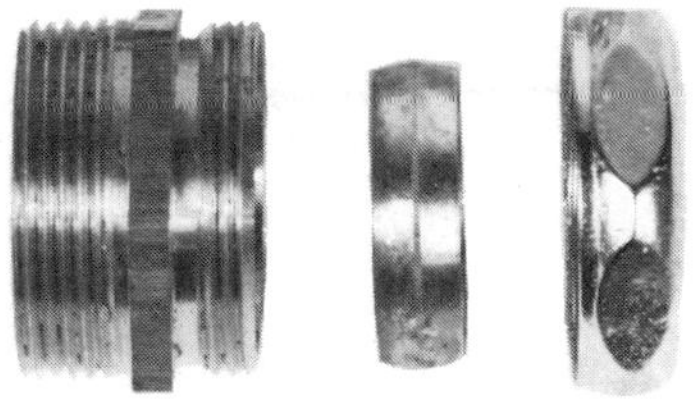

Fig. 8-2 P trap and compression trap adapter

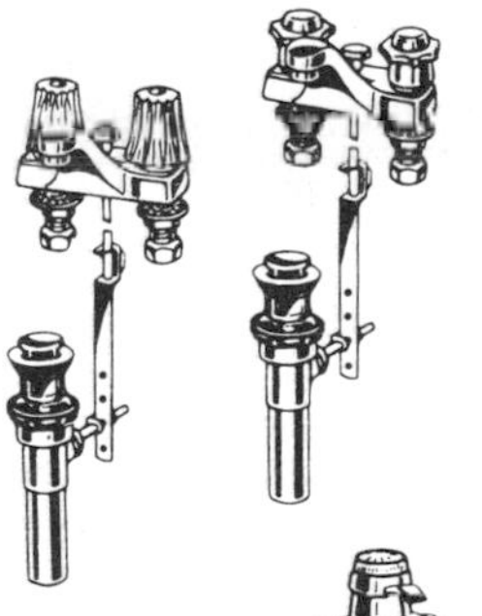

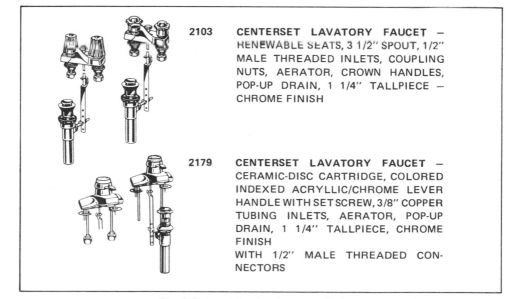

2103 CENTERSET LAVATORY FAUCET – RENEWABLE SEATS, 3 1/2" SPOUT, 1/2" MALE THREADED INLETS, COUPLING NUTS, AERATOR, CROWN HANDLES, POP-UP DRAIN, 1 1/4" TALLPIECE – CHROME FINISH

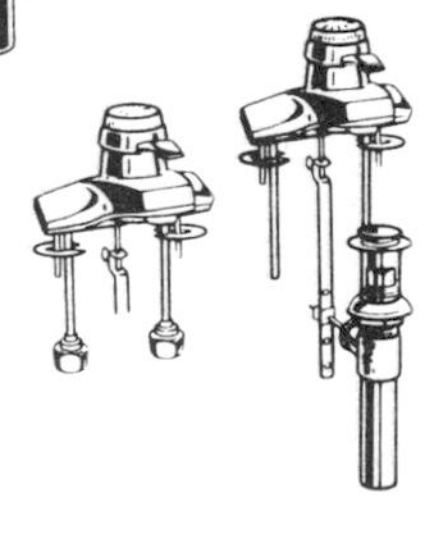

2179 CENTERSET LAVATORY FAUCET – CERAMIC-DISC CARTRIDGE, COLORED INDEXED ACRYLLIC/CHROME LEVER HANDLE WITH SETSCREW, 3/8" COPPER TUBING INLETS, AERATOR, POP-UP DRAIN, 1 1/4" TALLPIECE, CHROME FINISH
WITH 1/2" MALE THREADED CONNECTORS

Fig. 8-3 Combination lavatory fittings

2. How far above the finished floor is the centerline of the hanger?
   a. 32 1/4 inches   c. 31 inches
   b. 28 1/4 inches   d. 26 1/2 inches

3. What is the width of the lavatory against the wall?
   a. 18 inches   c. 10 1/2 inches
   b. 19 inches   d. 17 inches

4. What is the width of the hanger?
   a. 18 inches   c. 10 1/2 inches
   b. 19 inches   d. 8 inches

5. With a 3-inch clearance, how far is the centerline of the lavatory from the wall?
   a. 8 1/2 inches   c. 12 1/2 inches
   b. 12 inches   d. 7 inches

6. How far are the water supplies above the finished floor when using 2103 series fittings?
   a. 22 inches   c. 17 1/2 inches
   b. 20 3/4 inches   d. 24 1/2 inches

7. What is the spread of the water supplies?
   a. 10 1/2 inches   c. 2 1/4 inches
   b. 4 inches   d. 8 inches

8. What is the spread of the water lines at the connection to the faucet?
   a. 8 inches   c. 2 1/4 inches
   b. 4 inches   d. 10 1/2 inches

9. How much offset is needed for each bendable tube between the supply and faucet?
   a. 4 inches   c. 2 inches
   b. 8 inches   d. 1 1/4 inches

10. What is the standard pipe size for the faucet connection?
    a. 3/8 inch   c. 3/4 inch
    b. 1/2 inch   d. 1 1/4 inches

11. What is the standard pipe size for the water supplies?
    a. 3/8 inch   c. 3/4 inch
    b. 1/2 inch   d. 1 1/4 inches

12. What is the distance from the rim of the lavatory to the center of the waste opening?
    a. 13 1/2 inches   c. 14 3/4 inches
    b. 17 1/2 inches   d. 14 1/4 inches

13. What size trap is needed?
    a. 2″ x 1 1/2″   c. 1 1/4″ x 1 1/2″
    b. 1 1/2″ x 1 1/2″   d. 1 1/4″ x 1 1/4″

14. What is the distance from the center of the waste opening to the wall?
   a. 4 inches   c. 17 inches
   b. 2 1/4 inches   d. 6 1/2 inches

15. What is the distance from the center of the hot water faucet to the center of the waste control?
   a. 4 inches   c. 2 1/4 inches
   b. 2 inches   d. 9 inches

Unit 9 ROUGH-IN FOR WATER CLOSET COMBINATION

OBJECTIVES

After completing this unit, the student will be able to:

- discuss the installation of a water closet.
- determine locations of water supply and soil waste for water closet installation.

ONE- AND TWO-PIECE COMBINATIONS

The toilet combination consists of a flush tank and a bowl. Recent designs are freestanding with the tank supported on and fastened to the bowl. The freestanding design does not require backing, but a wall-supported tank does.

The two rough-in sheets in figures 9-1 and 9-2 show a two-piece combination and a one-piece combination. The two-piece construction bolts the tank to the bowl. The one-piece design has the bowl and tank cast in one piece. In all cases the rough-in gives necessary information to locate the soil waste

CADET TOILET

VITREOUS CHINA — CLOSE-COUPLED COMBINATION
SHOWN WITH
3/8 FLEX. SUPPLY

2122. SER.

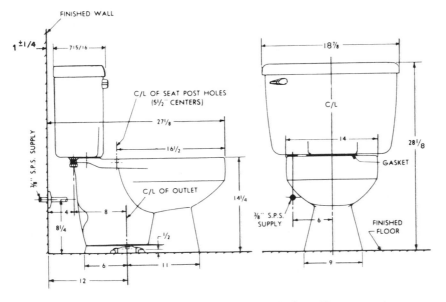

Fig. 9-1 Two-piece Cadet toilet

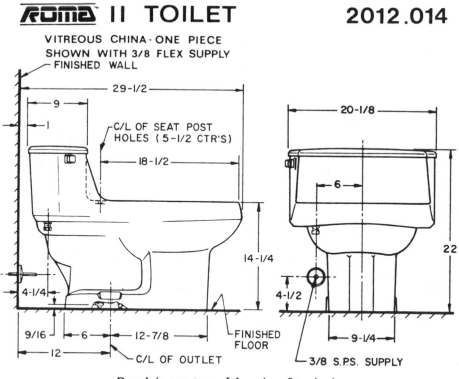

ꞓꞭ꞉ II TOILET   2012.014

VITREOUS CHINA - ONE PIECE
SHOWN WITH 3/8 FLEX SUPPLY

Rough-in courtesy of American Standard.
Fig. 9-2 One-piece Roma toilet

and the water supply. For public use with a flush tank, the two-piece combination requires the elongated bowl. The elongated bowl is shown in Appendix E, page 139.

The bowl is connected to the waste by bolting it through a floor flange attached to the waste pipe. The waste opening is 4 inches.

A gasket provides a watertight and gastight seal. Often other bolts attach to the floor for a more rigid installation.

The water supply is connected at the tank by a tube with a flanged end. The nut is tightened to compress a gasket between flanged end and the tank connection.

ASSIGNMENT

Multiple Choice

Compare the toilet combinations of the two rough-in sheets in figures 9-1 and 9-2 for likenesses and differences. Then read each question and circle the correct answer.

1. What is the minumum rough-in dimension from the finished wall to the center of the waste outlet for the Cadet toilet?
   a. 8 1/4 inches   c. 10 1/2 inches
   b. 7 3/4 inches   d. 12 inches

2. What is the minumum rough-in dimension from the finished wall to the center of the waste outlet for the Roma toilet?
   a. 2 5/8 inches   c. 12 inches
   b. 11 inches   d. 5 5/8 inches

3. What is the pipe size of the waste opening?
    a. 3/8 inch   c. 3 inches
    b. 1/2 inch   d. 4 inches

4. What is the distance from the finished wall to the front of the bowl for
    the elongated Cadet toilet?
    a. 28 inches   c. 27 1/8 inches
    b. 29 1/8 inches   d. 16 1/2 inches

5. What is the clearance between the finished wall and the tank for the
    Roma toilet?
    a. 3 1/4 inches   c. 4 inches
    b. 1 ± 1/4 inches   d. 1/2 inch

6. What is the clearance between the finished wall and the tank cover for
    the Cadet toilet?
    a. 3 1/4 inches   c. 4 inches
    b. 1 1/8 inches   d. 1/2 inch

7. If the wall finish is to be 1/2-inch wallboard and 3/8-inch tile, what is
    the minimum rough-in of the waste from the studs?
    a. 12 inches   c. 12 7/8 inches
    b. 12 1/2 inches   d. 12 3/8 inches

8. Using a 13-inch rough-in of the waste, how far will the front of the base
    of the Cadet toilet be from the finished wall?
    a. 24 inches   c. 18 1/4 inches
    b. 22 1/2 inches   d. 28 1/8 inches

9. What is the distance from the seat post holes to the front of the bowl
    for the elongated Cadet toilet?
    a. 16 1/2 inches   c. 29 1/8 inches
    b. 14 inches   d. 18 1/2 inches

10. The water supply for the Cadet toilet is how much higher than for the
    Roma?
    a. 2 5/8 inches   c. 8 1/4 inches
    b. 3 3/4 inches   d. 6 inches

11. How high is the top of the Roma toilet bowl from the finished floor?
    a. 19 9/16 inches   c. 14 1/4 inches
    b. 26 1/4 inches   d. 8 1/4 inches

12. What is the width of the elongated Cadet toilet bowl?
    a. 19 3/8 inches   c. 21 inches
    b. 14 inches   d. 9 inches

13. What is the difference in width between the Roma and the Cadet tanks?
    a. 1 1/4 inches   c. 1/4 inch
    b. 7 inches   d. 1/2 inch

14. If the rough-in of the waste is 17 1/2 inches to the side wall, what is the clearance between the side wall and the tank of the Cadet toilet?
    a. 10 1/2 inches   c. 7 13/16 inches
    b. 13 inches   d. 8 1/16 inches

15. What is the clearance if a Roma toilet is installed with a 17 1/2-inch rough-in to the side wall?
    a. 8 inches   c. 12 7/8 inches
    b. 7 7/16 inches   d. 9 inches

Unit 10 ROUGH-IN FOR BATHTUB
WITH SHOWER

OBJECTIVES

After completing this unit, the student will be able to:

- discuss the installation of a bathtub.
- determine locations for the tub waste and for the tub and shower water supply.

TUB SUPPORT

The built-in style tub rests on the floor and on supports at the wall. The wall support can be wood or an adjustable bracket bolted to a stud. Two or more brackets are needed for each tub.

WASTE OPENING IN FLOOR

A study of a tub waste rough-in will show that the waste boot is below the floor line. A 5″ x 10″ opening is cut out of the floor to accommodate this part of the waste connection. Slab-on-grade construction requires the same size depression in the concrete.

BACKING

The waste pipes are firmly supported by such wood backing as is necessary. The solid positioning of valves and piping makes the load of turning valves fall on the framework rather than on the wall plaster.

ACCESS DOOR

Good construction provides an access door to the waste connections and water piping of the tub so repairs can be made without damaging the wall. The access door is often in a concealed area, such as a closet.

ASSIGNMENT

Multiple Choice

The following questions are based on the rough-in sheets for the bathtub in figure 10-1, the bath drain in figure 10-2, page 36, and the shower combination in figure 10-3, page 37. Circle the correct answer for each question.

1. What is the greatest width of the tub?
   a. 28 inches
   b. 32 inches
   c. 30 1/2 inches
   d. 31 1/4 inches

2. What is the dimension from the finished wall to the centerline of the tub?
   a. 14 1/2 inches
   b. 15 1/4 inches
   c. 16 3/4 inches
   d. 14 inches

3. How far is the tub drain located from the studs of the end wall?
   a. 6 7/8 inches
   b. 1 3/8 inches
   c. 7 5/8 inches
   d. 3 inches

4. How far is the hot water supply roughed in from the studs of the side wall?
   a. 19 1/4 inches
   b. 11 1/4 inches
   c. 15 1/4 inches
   d. 20 inches

5. How far inside the rough wall construction is the cold-water riser located?
   a. 3 inches
   b. 1 3/8 inches
   c. 4 inches
   d. 2 1/4 inches

SPECTRA BATH

**ENAMELED CAST IRON - RECESS
SHOWN WITH 8620. SER. FITTING
& POP-UP CD & O**

2605. SER.
2607. SER.

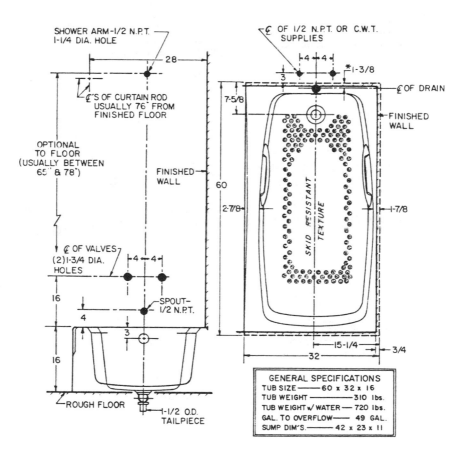

Rough-in courtesy of American Standard.

Fig. 10-1 Rough-in sheet for bathtub

6. The center line of the spout is how far above the rough floor line?
   a. 16 inches   c. 32 inches
   b. 20 inches   d. 28 inches

7. The water lines are what standard pipe size?
   a. 3/8 inch   c. 3/4 inch
   b. 1/2 inch   d. 1 1/2 inches

8. The spout is what size standard pipe thread?
   a. 3/8 inch   c. 3/4 inch
   b. 1/2 inch   d. 1 1/2 inches

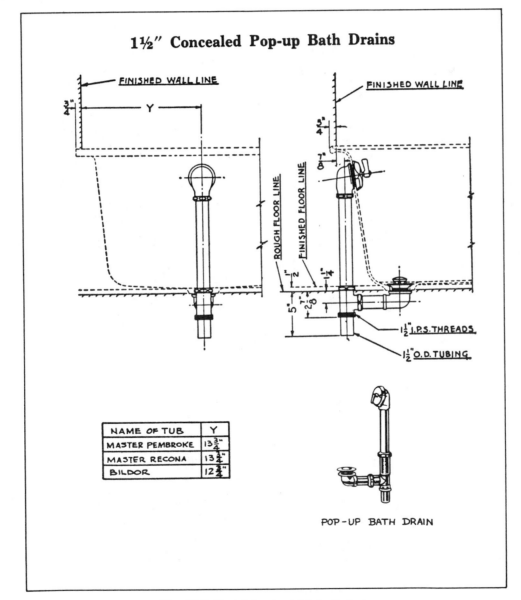

NAME OF TUB	Y
MASTER PEMBROKE	$13\frac{3}{4}"$
MASTER RECONA	$13\frac{1}{4}"$
BILDOR	$12\frac{1}{4}"$

POP-UP BATH DRAIN

Fig. 10-2 Rough-in sheet for bath drain

9. How far does the spout extend from the finished wall?
   a. 8 inches   c. 7 1/4 inches
   b. 5 inches   d. 3 1/8 inches

10. What is the greatest width inside the tub?
   a. 32 inches   c. 27 1/4 inches
   b. 28 1/4 inches   d. 26 1/4 inches

11. How far below the finished floor line is the horizontal center line of the drain?
   a. 5 inches   c. 1 3/4 inches
   b. 2 7/8 inches   d. 1 1/4 inches

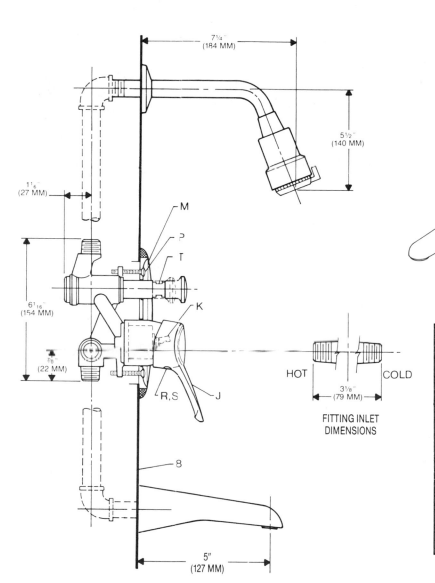

FIGURE 1

7¼"
(184 MM)

5½"
(140 MM)

1¹₁₆"
(27 MM)

6¹₁₆"
(154 MM)

⅞"
(22 MM)

M

P

T

K

R,S   J

8

HOT   COLD

3⅛"
(79 MM)

FITTING INLET
DIMENSIONS

5"
(127 MM)

Rough-in courtesy of American Standard.

Fig. 10-3 Single Control Bath/Shower Fitting With Built-in Diverter

12. How far is the tub waste centered from the rough end wall?
    a. 1 3/8 inches   c. 3 inches
    b. 1 5/8 inches   d. 7 5/8 inches

13. How far above the rough floor is the tub overflow opening?
    a. 15 1/4 inches   c. 13 inches
    b. 20 5/8 inches   d. 12 inches

14. What is the maximum recommended height of the shower arm above the rough floor?
    a. 6'-6"   c. 6'-0"
    b. 6'-4"   d. 6'-5"

15. How wide is the tub inside the finished wall?
    a. 31 1/4 inches   c. 30 1/8 inches
    b. 32 inches   d. 30 3/4 inches

16. How long is the tub inside the finished wall?
    a. 59 1/4 inches   c. 58 1/2 inches
    b. 60 inches   d. 57 3/4 inches

Isometric Pipe Drawings

Unit 11 FUNDAMENTALS OF ISOMETRIC DRAWING

OBJECTIVES

After completing this unit, the student will be able to:

- show why isometric drawings are used in the plumbing trade.
- describe isometric drawing of rectangular shapes.

PURPOSE OF THE ISOMETRIC DRAWING

The isometric drawing is a method of visualizing or showing a three-dimensional picture in one drawing. It is like a picture without many of the artistic details. Many plumbers have trouble visualizing a piping installation when working from a floor plan to an elevation and back again. The isometric drawing or sketch combines the plan and elevation views. It clearly shows the details of a piping installation.

The isometric system of drawing is used by the plumbing trade for two reasons: (1) it is easy to draw once the ideas are mastered, and (2) lines are drawn to scale length. It is possible to *scale* or measure lengths from an isometric drawing in the same manner that the architect's plan can be measured.

THE ISOMETRIC DRAWING

The isometric drawing follows certain rules in regard to directions in order to show three dimensions on a flat surface, figure 11-1.

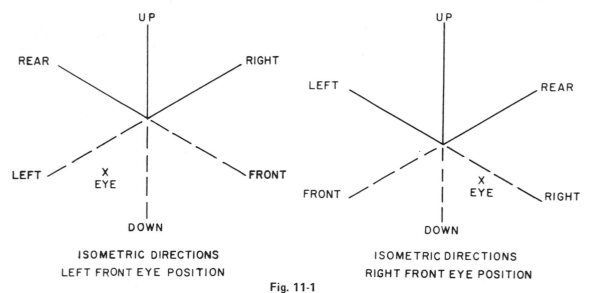

ISOMETRIC DIRECTIONS
LEFT FRONT EYE POSITION

ISOMETRIC DIRECTIONS
RIGHT FRONT EYE POSITION

Fig. 11-1

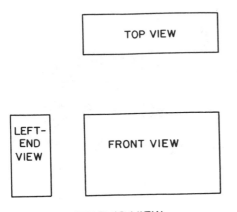

ORTHOGRAPHIC VIEW

Fig. 11-2

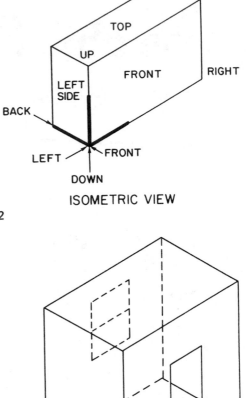

ISOMETRIC VIEW

These rules are:

- Lines that are vertical in an orthographic elevation remain vertical in the isometric sketch.

- Lines that are horizontal in an orthographic elevation are projected at an angle of 30 degrees in an isometric drawing.

COMPARING ISOMETRIC AND ORTHOGRAPHIC DRAWINGS

Compare the simple rectangular block in the three-view (orthographic) drawing on the left of figure 11-2, and the one-view (isometric) drawing on the right. Notice that:

- The vertical lines of the orthographic drawing remain vertical in the isometric.

- The horizontal lines of the orthographic drawing are not horizontal in the isometric but, instead, are projected at a 30-degree angle.

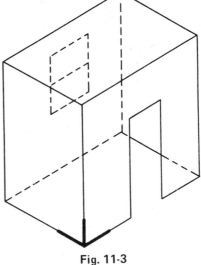

Fig. 11-3

- The lengths of the lines remain the same in the isometric as they were in the orthographic.

As soon as the drawing of the block is understood, the same idea can be applied to the drawing of the shape of a room, as in figure 11-3.

ASSIGNMENT

Review

Study isometric sketching using the explanation contained in Unit 11 of *Basic Construction Blueprint Reading.*

Drawing Exercises

1. Make an isometric drawing of a brick measuring 8' x 3 3/4" x 2 1/4". Make each line full size but do not dimension the drawing.

2. A bathroom is 5' wide, 8' long, and 7'-6" high. Make an isometric drawing of this shape to a scale of 1/2" = 1'-0". Lines should be drawn accurately but need not be dimensioned.

3. Practice similar projects until you have mastered the isometric representation of the rectangular solid.

Unit 12 AIDS TO DRAWING ISOMETRIC PIPE DRAWINGS

OBJECTIVES

After completing this unit, the student will be able to:

- visualize isometric pipe drawings.
- draw 45-degree diagonals in an isometric drawing.

INTERIOR VIEWS IN ISOMETRIC

In unit 11 the outside of a room was drawn. However, it is desirable to look into a room in order to visualize the piping layout. This can be done in one of several ways.

One method is to draw the room with fine light lines and the pipe drawing with heavier and darker lines, figure 12-1. This gives the effect of looking into a transparent room. This method requires drafting skills, however, and is difficult to do in field sketching.

Another way to visualize the piping layout is to section, or remove from the drawing, those parts which are in front of what is important to show. The usual section in a plumbing pipe layout leaves the ceiling and two walls out of the drawing. Figure 12-2 shows the steps in sectioning a room

A third method is even simpler as the room is shown only as a partial floor plan view. The walls are omitted from the drawing entirely. The walls are understood to be there but are left out so that the pipe drawing can be seen without unnecessary details, figure 12-3.

45-DEGREE ANGLES

The plumber thinks of a 45-degree offset as a diagonal in a square. Often the square is drawn to lay out a 45-degree angle. Such construction is shown in figure 12-4. A very similar method is used in isometric drawing to lay out a 45-degree angle.

In figure 12-5, the block has a 45-degree chamfer. It is located by measuring equal distances from the corner that would be there without the chamfer. A pipe drawing with a 45-degree diagonal, figure 12-6, page 44, would be very similar to the lines for part of the block as in figure 12-5.

To draw a 45-degree angle in isometric, begin with the 90-degree angle in isometric. Measuring an equal distance from the

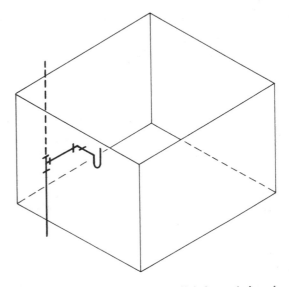

Fig. 12-1 Room lines are drawn lightly and the piping is outlined in heavier, darker lines.

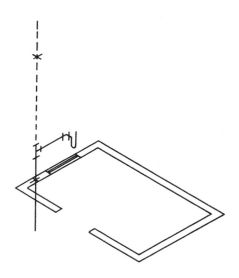

Fig. 12-3 Partial floor plan view

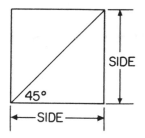

Fig. 12-4 A 45-degree line by diagonal in a square

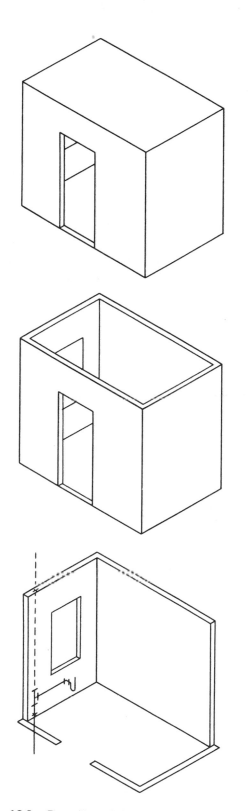

Fig. 12-2   Revealing piping by removing two walls
and the ceiling

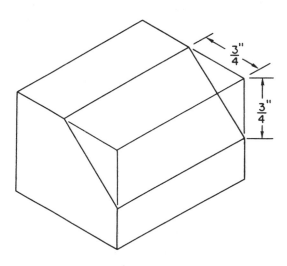

Fig. 12-5 Block with 45-degree chamfer

intersection of the two legs of the 90-degree angle establishes the two sides of a square. Connecting these two points establishes the diagonal which is a 45-degree angle.

In figure 12-6, point A would be the intersection of the two legs of a 90-degree angle. Measure an equal distance along each leg (3/4 inch is used here) and locate points B and C. Connect points B and C. This establishes the 45-degree offset.

Fig. 12-6 Pipe drawing with 45-degree diagonal

ASSIGNMENT

Review

Study angles in isometric. This is explained in Unit 12 of *Basic Construction Blueprint Reading.*

Drawing Exercises

1. Draw an isometric pipe drawing similar to figure 12-7. Make all fittings 45 degrees instead of 90 degrees as shown.

2. Draw a room 8'-0" x 10'-0" x 7'-6" in an isometric representation with ceiling and two walls removed. Use a scale of 1/4" = 1'-0". Show the walls as 6 inches thick.

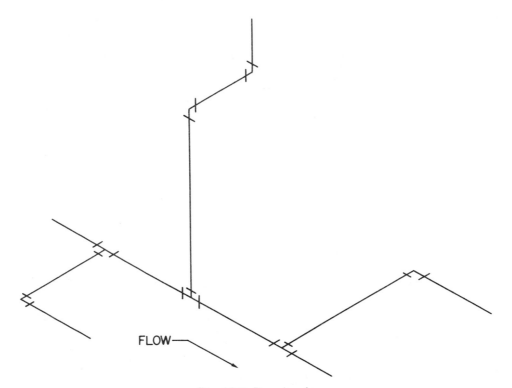

FLOW

Fig. 12-7 Pipe drawing

Unit 13 DIMENSIONING ISOMETRIC DRAWINGS

OBJECTIVES

After completing this unit, the student will able to:

- dimension an isometric drawing.
- show pipe size on an isometric pipe drawing.

DIMENSIONING AN ISOMETRIC DRAWING

An isometric drawing is dimensioned with extension lines and dimension lines in much the same manner as a two-dimensional drawing. The extension lines extend from the drawing so that the dimension line is parallel to the object line and of equal length to it.

It is somewhat more difficult to dimension the isometric drawing. There is only a single view and, therefore, less room is available than on three separate views. Figure 13-1 is a dimensioned isometric drawing for part of a pipe hanger.

DIMENSIONING AN ISOMETRIC PIPE DRAWING

Simplified dimensioning is best for an isometric pipe drawing. Because few dimensions are shown, it is important to scale

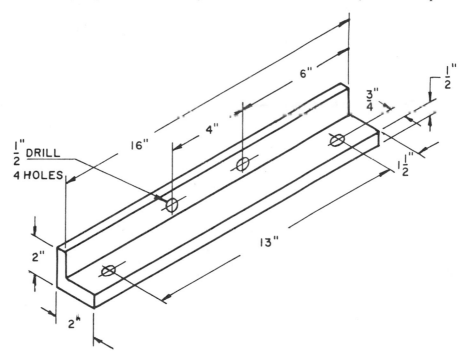

Fig. 13-1 Dimensioning a pipe hanger

accurately so that the drawing can be measured. Refer to the rough-in sheets and the architect's plans for the necessary information to make the isometric pipe drawing.

Since pipe drawings are measured from the center of one fitting to the center of the next fitting, it is possible to omit the extension and dimension lines by using a notation, such as *28'' c-c.*

Pipe sizes must be added to the pipe drawing. The size of a pipe is indicated by a number near the pipe, figure 13-2.

The plumbing code requires that the vent return be located higher than the rim of the fixture. In figure 13-3, note that the vent return is 48 inches above the floor. This is a good design.

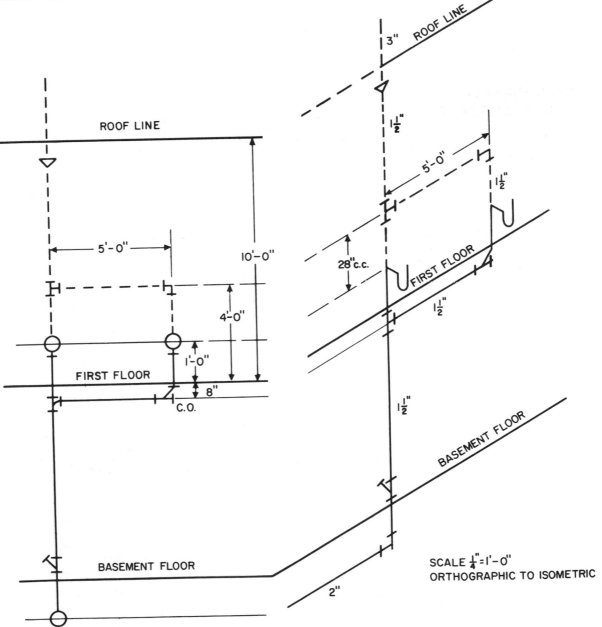

Fig. 13-2

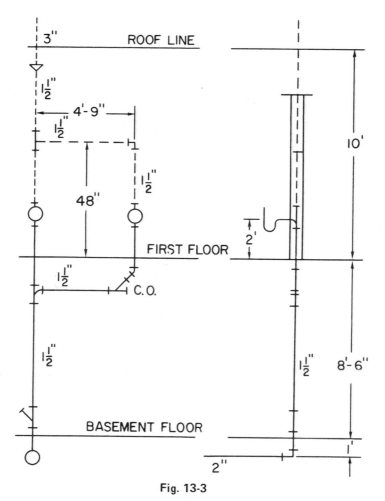

Fig. 13-3

ASSIGNMENT

Review

Review dimensioning isometric drawings. This is explained in Unit 13 of *Basic Construction Blueprint Reading.*

Drawing Exercise

Make an isometric pipe drawing of figure 13-3. Note that figure 13-3 is a front and right-hand view of one pipe layout. Study figure 13-2. Figure 13-2 is a similar pipe layout shown in both orthographic and isometric with the dimension lines carried from one to the other. List the materials needed.

Unit 14 PLUMBING CODES FOR WASTE AND VENT PIPING

OBJECTIVES

After completing this unit, the student will be able to:

- recognize the wording and tables of a plumbing code.
- use a plumbing code in blueprint reading and sketching.

PLUMBING CODES

The appendix includes paragraphs selected from the BOCA National Plumbing Code regulating waste and vent piping. The provisions apply to one- and two-family houses and to commercial buildings such as those studied in the text. The code paragraphs govern the sizing of waste and vent piping and permissible pipe fixtures. State and local codes have similar requirements and may have provisions for local conditions. Paragraph **P-101.2 Materials not provided for**, shown below, covers this situation.

This unit only introduces how plumbing codes are worded and how they affect a plumbing installation. For a more complete understanding of plumbing codes, additional study and experience is required.

P-101.2 Matters not provided for: Any plumbing requirement essential for the sanitary safety of an existing or proposed building or structure or essential for the safety of the occupants thereof, and which is not specifically covered by this code, shall be determined by the code official.*

P-403.3.2 Drainage pipe in filled ground: When a building sewer or building drain is installed on filled or unstable ground, the drainage pipe shall conform to one of the standards for ABS plastic pipe, cast iron pipe, copper tube, or PVC plastic pipe listed in Table P-403.3.2.*

Table P-403.3.2
BUILDING SEWER PIPE*

Material	Standard (see Appendix A)
Acrylonitrile butadiene styrene (ABS) plastic pipe[a]	ASTM[b] D2661; ASTM D2751; ASTM F628
Asbestos cement pipe	ASTM C428; ASTM C644
Bituminized fiber pipe	ASTM D1861; ASTM D1862
Cast iron pipe	ASTM A74; CISPI 301
Concrete pipe	ASTM C14; ASTM C76
Copper tubing (Type K or L)	ASTM B75; ASTM B88; ASTM B251
Polyvinyl chloride (PVC) plastic pipe[a]	ASTM D2665; ASTM D2949; ASTM D3033; ASTM D3034
Vitrified clay pipe	ASTM C4; ASTM C700

Note a: Thermoplastic sewer pipe shall be installed in accordance with ASTM D2321 listed in Appendix A.
Note b: American Society for Testing Materials.
*BOCA National Plumbing Code/1987, 7th Edition, Copyright 1986, Building Officials and Code Administrators International, Inc. Published by arrangements with author. All rights reserved. No parts of this book may be reproduced or transmitted in any form or by any means, elelctronic or mechanical, including photocopying, recording or by an information storage and retrieval system without advance permission in writing from Building Officials and Code Administrators International, Inc. For information, address: BOCA International, Inc., 4051 West Flossmoor Road, Country Club Hills, IL 60477.

ASSIGNMENT

Multiple Choice

Study the paragraphs of the *BOCA National Plumbing Code* in the appendix. The following questions are based on these codes. Circle the correct answer for each one.

1. What type of toilet bowl is required for public use?
   a. Wall hung
   b. Elongated bowl
   c. One-piece tank and bowl
   d. Round front bowl

2. How is the installation of plumbing sewers regulated?
   a. By the Plumbing Code.
   b. By the local authorities having jurisdiction.
   c. By the engineering design.
   d. By the working drawings.

3. What slope is required for a horizontal 2-inch waste?
   a. 1/2 inch per foot
   b. 1/4 inch per foot
   c. 1/8 inch per foot
   d. 1/16 inch per foot

4. What slope is required for a horizontal 4-inch waste?
    a. 1/2 inch per foot   c. 1/8 inch per foot
    b. 1/4 inch per foot   d. 1/16 inch per foot

5. Which of the following fittings cannot be used when changing the direction of a drainage pipe?
    a. Tees   c. 1/8-inch bends
    b. 45-degree wyes   d. Long sweeps

6. How is the total discharge flow of a fixture measured?
    a. In cubic feet per second   c. In gallons per minute
    b. In gallons per second   d. In cubic feet per minute

7. What is the drainage unit requirement for a public water closet?
    a. 6   c. 4
    b. 8   d. 2

8. How many fixture units may be connected to a 2-inch horizontal branch drain?
    a. 6   c. 24
    b. 10   d. 26

9. How many flush tank water closets may be connected to a 3-inch stack?
    a. 20   c. 30
    b. 2   d. 6

10. How small may underground waste piping be?
    a. 1 1/4 inches   c. 2 inches
    b. 1 1/2 inches   d. 2 1/2 inches

11. What is the smallest permissible vent pipe extension through the roof?
    a. 3 inches   c. 2 1/2 inches
    b. 2 inches   d. 1 1/4 inches

12. What is the greatest permissible distance for a 3-inch fixture drain from trap to vent?
    a. 3 1/2 feet   c. 6 feet
    b. 5 feet   d. 10 feet

13. What is the smallest size pipe permitted for vent use?
    a. 1 1/4 inches   c. 2 inches
    b. 1 1/2 inches   d. 2 1/2 inches

14. How many drainage units may be connected to a 3-inch waste stack?
    a. 42   c. 53
    b. 102   d. 320

15. What is the maximum length for a 2-inch vent stack and branch vents?
    a. 24 feet   c. 150 feet
    b. 50 feet   d. 200 feet

Unit 15 WASTE AND VENT FOR KITCHEN SINK

OBJECTIVES

After completing this unit, the student will be able to:

- make a waste and vent isometric pipe drawing for a kitchen sink.

- list the fittings and pipe for waste and vent lines to the kitchen sink.

PIPING FOR KITCHEN SINK

When a sink is installed under a window, the waste stack must be at one side of the window, figure 15-1.

The sink trap connects into a drainage elbow and through a horizontal pipe to a TY fitting in the waste stack. The stack above the TY is the vent and is shown by a broken line. Below the TY is the waste. It is shown by a solid line.

The waste line pipe size usually required by the plumbing code is 1 1/2 inches, although a 1 1/4-inch vent may be used. The vent must increase one pipe size before going through the roof. This is to prevent hoar frost from closing the vent. The vent then continues 24 inches above the roof.

A cleanout should be provided by installing a fitting in the waste line above the cellar floor. The opening for the cleanout is closed by a pipe plug.

The house drain under the basement floor should be the pipe material and size specified in the plumbing code. For an example, refer to paragraphs P 403.3.2 on page 48 and P 602.1 in the appendix.

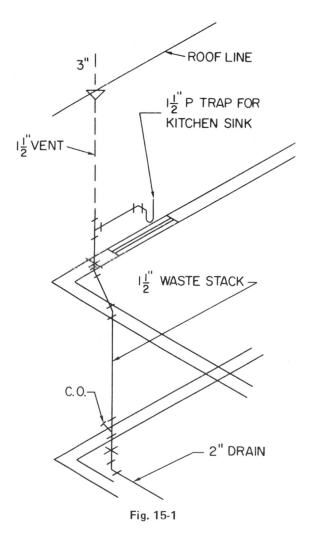

Fig. 15-1

ASSIGNMENT

Drawing Exercise

Study the sink plan shown in figure 15-2. Make an isometric pipe drawing for the waste and vent. Start the drawing as suggested in figure 15-3. Indicate clearly the pipe size and material. Make a list of fittings and pipe needed to complete the stack.

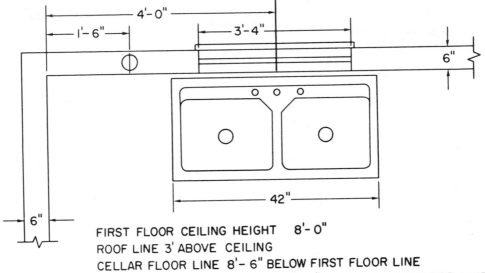

FIRST FLOOR CEILING HEIGHT   8'- 0"
ROOF LINE 3' ABOVE CEILING
CELLAR FLOOR LINE 8'- 6" BELOW FIRST FLOOR LINE
CENTERLINE OF HOUSE DRAIN 1'- 0" BELOW CELLAR FLOOR LINE
KITCHEN WASTE TO ROUGH-IN 20" ABOVE FIRST FLOOR LINE

Fig. 15-2 Kitchen sink plan

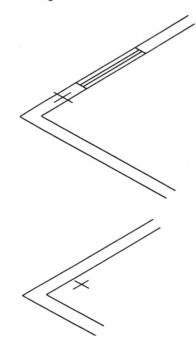

Fig. 15-3

Unit 16 WASTE AND VENT FOR
TWO LAVATORIES

OBJECTIVES

After completing this unit, the student will be able to:

- plan waste and vent piping by using an isometric pipe drawing.
- list the fittings and pipe for waste and vent lines to the lavatory.

PIPING FOR TWO LAVATORIES

The pipe size of the waste stack for two sinks or lavatories is determined by the plumbing code.

Each horizontal run from lavatory to stack can be 1 1/2-inch pipe and by some codes must not be longer than 42 inches without a separate vent, figure 16-1. A cleanout should be installed above the cellar floor. The house drain under the cellar floor should be at least 2-inch pipe.

The vent stack is often required to increase one pipe size before projecting through the roof. A minimum size of 3 inches can be required.

The new fitting in the assembly is a double TY. This is a drainage fitting with connections on both sides. Pipe sizes are determined by plumbing codes.

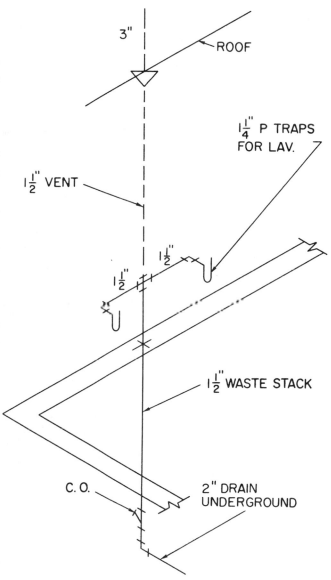

Fig. 16-1 Waste and vent for two lavatories

ASSIGNMENT

Drawing Exercise

Note: Dimensions have been omitted in figure 16-2, and in additional figures in subsequent units, to give the student practice in determining the dimensions from information given on the figure. The student should fill in all missing dimensions as part of the Assignment.

1. Draw an isometric pipe drawing for the waste and vent of two lavatories as shown in the plan view, figure 16-2. Use a solid line for the waste and a broken line for the vent. This drawing will be similar to the one for the kitchen sink in Unit 15. Show the pipe size and material, including lavatory traps.

2. Write a list of fittings and length of pipe to complete the waste and vent assembly in figure 16-2.

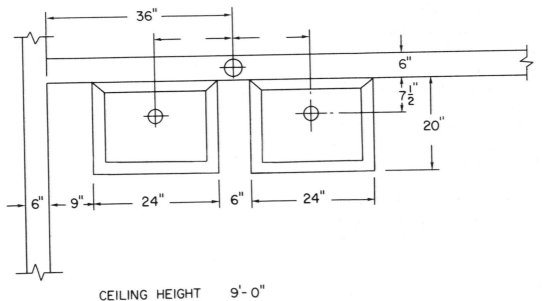

CEILING HEIGHT   9'- 0"

ROOF LINE 15" ABOVE CEILING

CELLAR FLOOR LINE 8'- 6" BELOW FIRST FLOOR LINE

HOUSE DRAIN 1'- 0" BELOW CELLAR FLOOR LINE

ROUGH IN WASTES 18" ABOVE FIRST FLOOR LINE

Fig. 16-2 Two lavatories

Unit 17 THREE LAVATORIES TO A SINGLE STACK

OBJECTIVES

After completing this unit, the student will be able to:

- discuss the connection of waste and vent piping, including a back vent.
- make an isometric sketch for future reference.

HORIZONTAL WASTE PIPE WITHOUT ADDITIONAL VENT

The length of an unvented horizontal run from a trap connection to the stack is limited by plumbing codes and good design. In this text, a length no longer than 3'-6" is used without venting the fixture.

THE BACK VENT

Distant traps use a back-vented system. The trap connects to a TY fitting from which waste piping goes down and vent piping upward. Both waste and vent piping turn and are connected at separate points to the stack. Usually the waste line runs under the floor. The vent should rise before turning to connect to the stack.

Figure 17-1 is an elevation sketch of waste and vent piping including a back vent. "A" is a connection for a sink or lavatory close to the stack (known as high fixture). "B" shows a back vent connected to a fixture roughing which is more than 3 1/2 feet from the main stack.

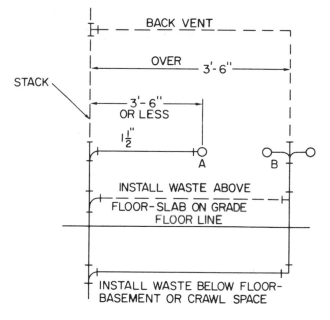

Fig. 17-1 High fixture and back vent

ASSIGNMENT

Drawing Exercises

1. Figure 17-2, page 56, shows three lavatories in line. Make an isometric pipe drawing and include a back-vented system for the far lavatory.

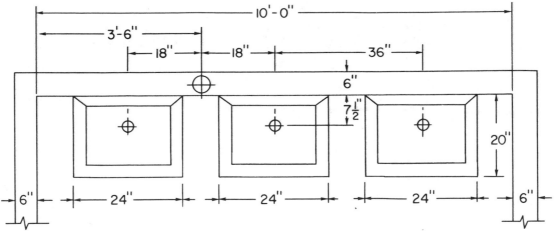

CEILING HEIGHT 8'-3"

ROOF LINE 4'-6" ABOVE CEILING LINE
BASEMENT FLOOR 9'-0" BELOW FIRST FLOOR LINE
HOUSE DRAIN 2'-0" BELOW BASEMENT FLOOR LINE
LAVATORY WASTES ROUGH-IN 18" ABOVE FLOOR

NOTE: ABOVE DIMENSIONS APPLY TO FIGS. 17-2, 17-3, AND 17-4.

Fig. 17-2 Three lavatories in line on inside partition

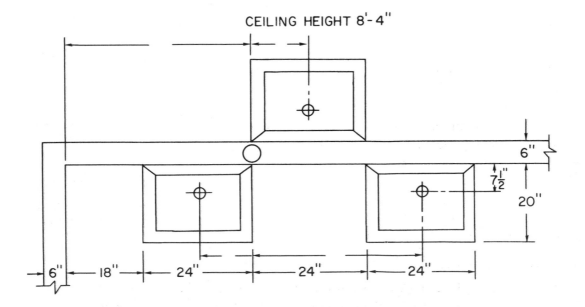

CEILING HEIGHT 8'-4"

Fig. 17-3 Variation of three in line lavatories

2. Study figure 17-3 and make an isometric pipe drawing. Include the variation of the lavatory on the opposite side of the partition. Fill in the missing dimensions in figure 17-3.

3. Study figure 17-4. Fill in the missing dimensions. Make an isometric pipe drawing for the waste and vent for two lavatories and a service or slop sink in three different rooms.

4. In the space provided, make a sketch for reference of the isometric pipe drawing of figure 17-2.

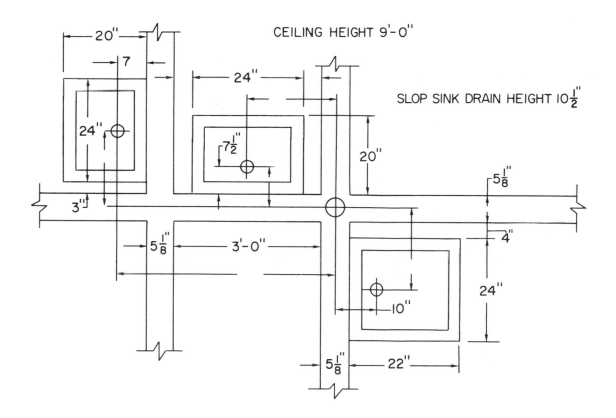

Fig. 17-4 Two lavatories and slop sink to single stack

ISOMETRIC SKETCH OF PIPING
OF FIGURE 17-2

Unit 18 SOIL STACK, WASTE AND VENT PIPING FOR WATER CLOSET AND LAVATORY

OBJECTIVES

After completing this unit, the student will be able to:

- discuss the waste and vent piping for a water closet.
- plan an installation with a water closet and a lavatory.
- sketch an isometric pipe drawing for future reference.

PIPE MATERIALS

There are different pipe materials in use. PVC and ABS pipe give economy of pipe and fitting costs, as well as labor cost. Copper pipe can be specified. The use of a cast iron waste and vent stack with a loop made of threaded pipe requires fittings with special design not needed for other materials. The drawings and blueprint reading are the same for all. The materials used changes visualization for the assembly of the pipe.

THE CLOSET BEND

The waste connection to a water closet is made by installing a TY connection in the stack below the floor line. A closet bend is connected into the side opening of the TY.

The closet bend is a 90-degree fitting or quarter bend with one leg longer than the other, figure 18-1. The shorter leg comes through the opening in the floor. By various approved methods, the closet bend is connected to the closet bowl. The minimum size soil stack is 3 inches.

THE TAPPED TEE

A threaded pipe vent can connect to a cast iron stack by a tapped tee fitting, figure

Fig. 18-1 Closet bends

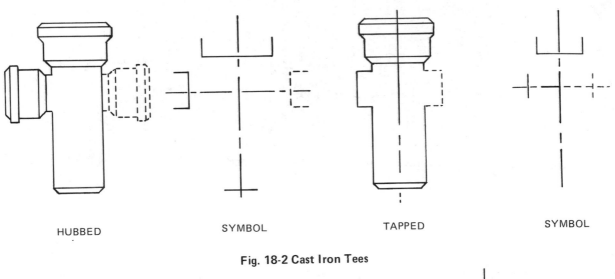

HUBBED   SYMBOL   TAPPED   SYMBOL

Fig. 18-2 Cast Iron Tees

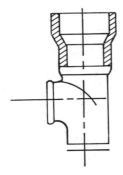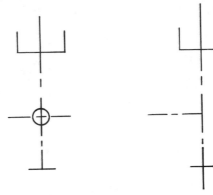

SYMBOLS

Fig. 18-3 Tucker or Slip and Caulk TY

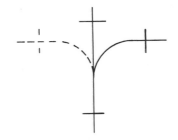

Fig. 18-4 Single and Double TYs (Sanitary Tees)

18-2, page 60. A bushing can be used to reduce the 2-inch opening to 1 1/2- or 1 1/4-inch pipe size.

THE TUCKER HUB TY

With threaded pipe some means is needed to make the last connection in the loop formed by the lavatory waste and vent piping. A TY with a special hub connection can be used. The hub is made tight to steel pipe with lead and oakum, or equal. See figure 18-3, page 60. *Note:* a union is not permitted.

INCREASER FOR VENTS

The vent stack can be completed with an increaser and stack length in one piece. The plumbing codes often require stacks under

Fig. 18-5 Quarter bend with side inlet

4-inch size to increase one pipe size before going through the roof. Some states specify that for public buildings all stacks must increase to 4-inch size before going through the roof.

STACK VENTING THE WATER CLOSET

When the water closet is connected to the soil stack as the high fixture, no additional venting is required. The water closet is stack-vented by this installation. Other fixtures, such as a lavatory, are connected into the stack by a sanitary tee installed below the TY for the water closet, figure 18-4.

WET VENTING THE WATER CLOSET

Closet bends can be purchased with two side openings. Other fixtures can be connected to the soil stack at the side openings. These can be either 1 1/2-inch or 2-inch threaded pipe size. In copper or plastic, the closet bends can be made up by using side outlet quarter bends, figure 18-5.

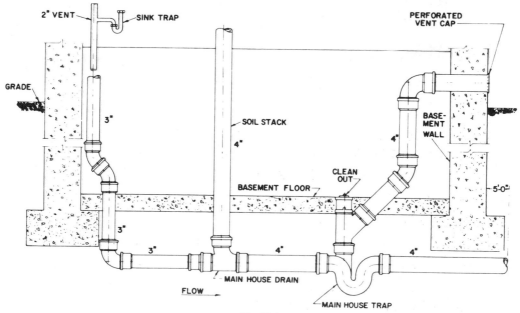

Fig. 18-6

When a lavatory, for instance, is connected to the side opening of the closet bend, the lavatory should be back-vented to the stack, as shown in Unit 17. The water closet then vents through the lavatory piping. This is called a wet vent, as the lavatory waste lines serve two purposes: to drain the lavatory and to vent the water closet. The wet vent is 1 1/2-inch pipe for a single fixture.

HOUSE DRAIN SIZE

The house drain to the base of the soil stack may be 4-inch size; however, current codes may permit the use of 3-inch pipe, figure 18-6. Branches from the house drain to the base of the waste stack can be smaller size. The waste stack does not carry the discharge from a water closet.

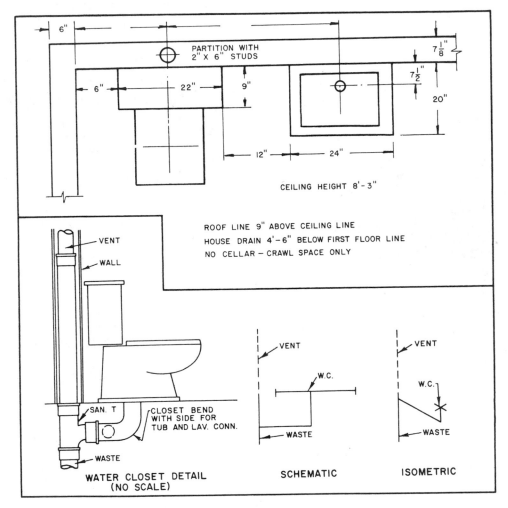

Fig. 18-7

ASSIGNMENT

Drawing Exercises

1. Study the plan and notes in figure 18-7. Fill in the missing dimensions. Make an isometric pipe drawing of the waste and vent piping.

2. In the space provided, make an isometric sketch of the waste and vent piping for figure 18-7. This is for future reference.

ISOMETRIC SKETCH OF PIPING
OF FIGURE 18 – 7

Unit 19 THREE-FIXTURE BATH ON ONE WALL

OBJECTIVES

After completing this unit, the student will be able to:

- discuss soil stack and waste and vent piping for a three-fixture bathroom.
- recognize where pipe sizes must be increased.
- sketch the waste and vent system for future reference.

FIXTURES ROUGHED ON ONE WALL

Installing all the piping for a bathroom on one wall is economical. The lengths of pipe are shorter and less labor is needed for the installation. It is possible to have a well-arranged bathroom that is roughed on one wall.

There are two floor plans that are possible. In the first plan the water closet is the center fixture, figure 19-1. In the second plan the water closet is an end fixture, figure 19-2. In either case, the stack is located close to the water closet, and the lavatory and tub are back-vented to the stack.

PIPE SIZES

The stack must be at least 3-inch size. The tub or lavatory requires 1 1/2-inch pipe when the waste lines run separately to the stack, as in a bathroom with the water closet as a center fixture.

When the waste from the tub and lavatory join in one pipe, a 2-inch waste is required for the combined flow. This is the case where the water closet is the end fixture in the bathroom.

WATER CLOSET VENTING

The water closet can be stack-vented when connected as a high fixture. The tub and lavatory waste connects to the stack through a fitting below the TY for the water closet or to an opening in the closet bend.

It is also general practice, in some areas, to connect the lavatory and the tub waste to the side opening closet bend. In this case, the water closet is wet-vented through the piping for the other fixture. The horizontal vent line for this fixture is increased to a 2-inch size when designed to serve as a wet vent.

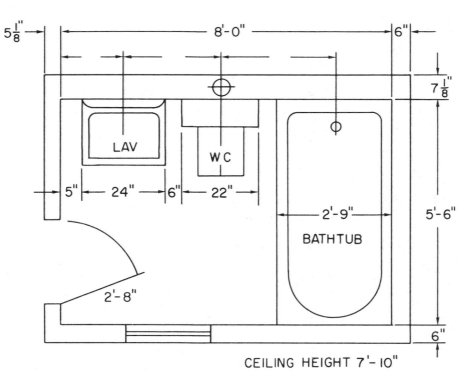

Fig. 19-1

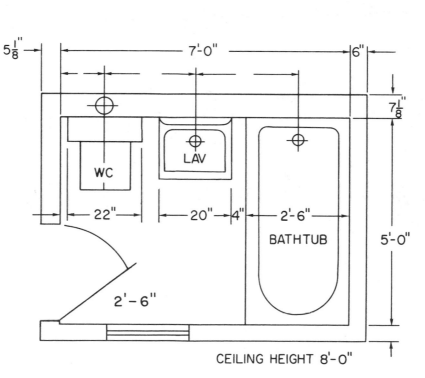

FIRST FLOOR BATHROOMS- USE P TRAP IN TUB WASTE
ROOF LINES 9'-6" ABOVE FIRST FLOOR LINE
HOUSE DRAIN 9'-0" BELOW FIRST FLOOR LINE
CELLAR FLOOR 8'-0" BELOW FIRST FLOOR LINE

Fig. 19-2

ASSIGNMENT

Drawing Exercises

1. Study figure 19-3. Make an isometric piping drawing to show a satisfactory waste and vent system for the bath in figure 19-1. Connect the waste for lavatory and tub through the closet bend.

2. Study figure 19-4 and make an isometric pipe drawing for waste and vent system for the bath in figure 19-2. Connect the waste for tub and lavatory to a separate TY in the stack.

3. Make a sketch of the approved drawing for figure 19-2 for future reference. Sketch it in the space provided.

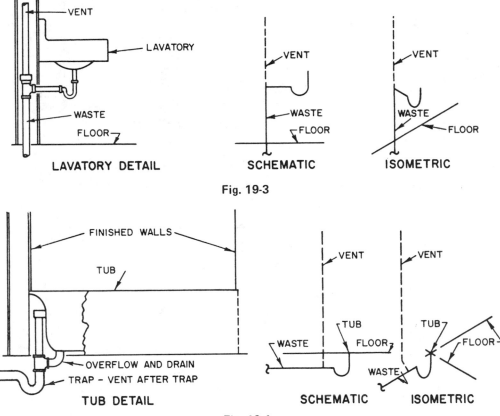

Fig. 19-3

Fig. 19-4

ISOMETRIC SKETCH OF PIPING
OF FIGURE 19-2

Unit 20 BATHROOM FIXTURES
ON OPPOSITE WALLS

OBJECTIVES

After completing this unit, the student will be able to:

- discuss the design and location of waste and vent piping with bathroom fixtures on opposite walls.
- plan and draw vent piping.

PIPE SIZE

The longer lengths of pipe used for bathroom fixtures on opposite walls require the same size pipe as those needed for the bath roughed in one wall. The fittings used are also of the same kind.

FIRST AND SECOND FLOOR BATHS

The first floor bath is easier to plan and install because the waste piping can be run under the floor joists. The second floor bath requires that pipe be run either between or through the joists.

The P trap, figure 20-1, is often used in the tub waste on a first floor bath as it is accessible and may be cleaned from the bottom.

The drum trap, figure 20-2, is used for the waste of a second floor tub. The trap cover is flush with the floor and can be removed to clean the trap. Some codes require drum traps for all tubs, and some installations on the first floor have the drum trap with the cover facing down.

In all cases, the vent piping must rise and therefore needs to be installed in the wall. A vent line that does not rise would be choked by water and unable to supply air to balance pressures in the pipe. Whenever possible, the piping should be in inside partitions rather than in outside walls. There is less chance of freezing this way. It is also easier to install insulation in walls when the dead air spaces are free from obstructions.

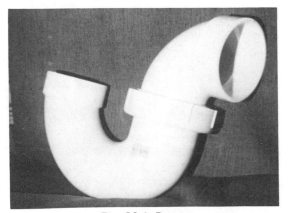

Fig. 20-1 P trap

Fig. 20-2 Drum traps

VENT PIPING OVER DOORS

In making an isometric pipe drawing, it is sometimes desirable to show a wall in addition to the floor plan. When the vent pipe is in a partition with a door, planning can be more accurate if the partition with the door opening and the position of the vent pipe are all shown.

ASSIGNMENT

Drawing Exercise

Study figure 20-3. Design a satisfactory waste and vent by means of an isometric pipe drawing. Show the partition with the door and run the tub vent in this partition. Connect the waste for the tub and lavatory into a double side-opening closet bend.

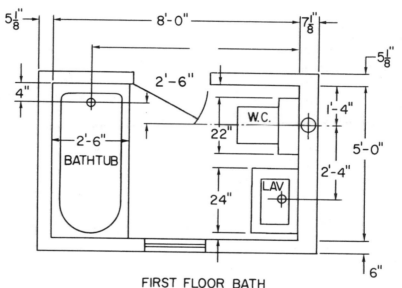

FIRST FLOOR BATH
ROOF LINE 3'-9" ABOVE CEILING LINE
BASEMENT FLOOR 8'-0" BELOW FIRST FLOOR
HOUSE DRAIN 8'-8" BELOW FIRST FLOOR

Fig. 20-3

Unit 21 FOUR-FIXTURE BATH

OBJECTIVES

After completing this unit, the student will be able to:

- discuss the waste and vent piping for a four-fixture bath.

- make an isometric sketch for future reference.

FOUR-FIXTURE BATH

The four-fixture bathroom may use a separate shower or a second lavatory as the additional fixture. A shower or lavatory as a fourth fixture requires quite similar waste and vent systems. Either fixture needs a 1 1/2-inch waste and vent. As before, the combined waste or vent for two 1 1/2-inch lines is carried by a 2-inch pipe.

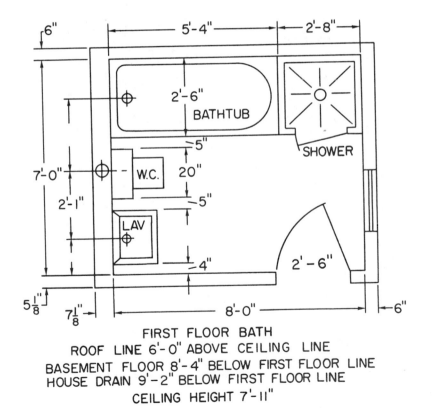

FIRST FLOOR BATH
ROOF LINE 6'-0" ABOVE CEILING LINE
BASEMENT FLOOR 8'-4" BELOW FIRST FLOOR LINE
HOUSE DRAIN 9'-2" BELOW FIRST FLOOR LINE
CEILING HEIGHT 7'-11"

Fig. 21-1

ASSIGNMENT

Drawing Exercises

1. Make a neat isometric pipe drawing for waste and vent lines to convenient scale for the bathroom in figure 21-1.

2. Make an isometric sketch of the piping in figure 21-1 for future reference in the space provided.

ISOMETRIC SKETCH OF PIPING

OF FIGURE 21-1

Unit 22 TWO BATHROOMS: BACK TO BACK

OBJECTIVES

After completing this unit, the student will be able to:

- describe the arrangement of waste and vent piping for back-to-back bathrooms.

- make an isometric sketch for future reference.

BACK-TO-BACK BATHROOMS

Two bathrooms can be roughed on one wall. This is an economy feature used in homes, hotels, and other construction. It is similar to a single bath roughed on one wall except that the pipe size must be adequate for the combined flow or vent.

THE SOIL STACK

The soil stack can be a 3-inch size, the same as for a single bath. The water closets connect to the stack through a double TY, figure 22-1. As in other bathroom piping, the closets may be stack-vented or they may be wet-vented.

LAVATORY AND TUB WASTE AND VENTS

The waste lines and the vent lines for two lavatories or for two tubs need to be 2 inches in size. Sometimes the waste piping is separate for each fixture, while the vents are combined. This requires four 1 1/2-inch waste lines and 2-inch vent piping.

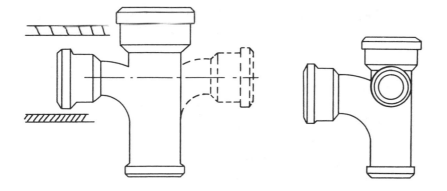

Fig. 22-1

ASSIGNMENT

Drawing Exercises

1. Draw accurately to scale an isometric pipe drawing for the back-to-back bathrooms in figure 22-2.

2. Draw an isometric sketch of the pipe drawing for future reference in the space provided.

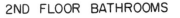

2ND FLOOR BATHROOMS

IST FL. CEILING 8'-2"

2ND FL. CEILING 7'-9"

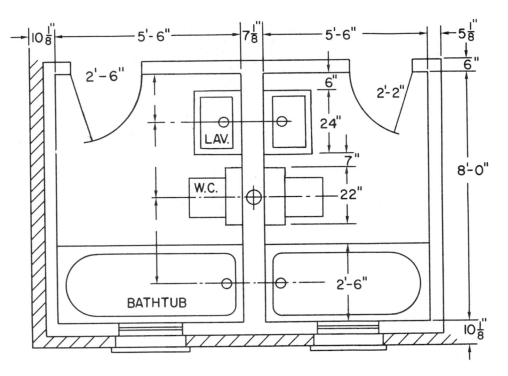

NOTE: BATHTUBS SHOULD NEVER BE PUT IN FRONT OF
WINDOWS EXCEPT IN REPAIR OR ALTERATION WORK

ROOF LINE 3'-6" ABOVE 2ND FLOOR CEILING

BASEMENT FLOOR 8'-0" BELOW FIRST FLOOR LINE

HOUSE DRAIN 9'-0" BELOW FIRST FLOOR LINE

Fig. 22-2

ISOMETRIC SKETCH OF PIPING
OF FIGURE 22-2

Unit 23 BATHROOM AND KITCHEN FIXTURES INTO ONE STACK

OBJECTIVES

After completing this unit, the student will be able to:

- discuss the waste and vent piping for back-to-back installation of bath and kitchen.

- make an isometric sketch for future reference.

The back-to-back designs are economy features in plumbing because only one stack is used. Often it is desirable to place the kitchen sink on an outside wall under a window. This requires longer waste and vent piping than the back-to-back position, but is similar in other respects.

PIPE SIZE

The stack must be of 3-inch size. The water closet can be connected for stack venting or for wet venting, although the wet vent is more often used.

The waste and the vent piping from each of the other fixtures can be 1 1/2-inch size. When a pipe carries the waste or vent from two of these fixtures, however, it must be of the 2-inch size.

ASSIGNMENT

Drawing Exercises

1. Make an isometric pipe drawing to convenient scale for the back-to-back bath and kitchen in figure 23-1. Use a wet-vent system.

2. Sketch an isometric pipe drawing for future reference in the space provided.

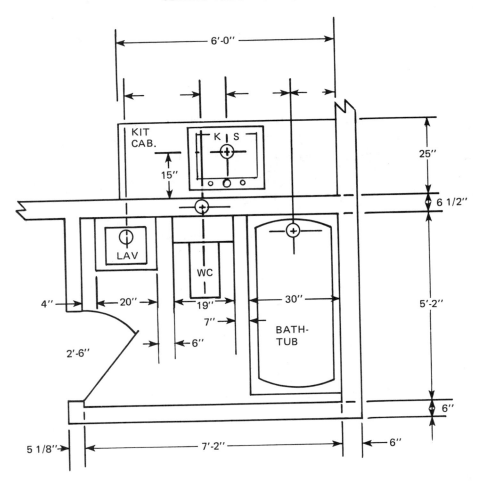

FIRST FLOOR PLAN

ROOF LINE 4'-6" ABOVE CEILING LINE

BASEMENT FLOOR LINE 8'-0" BELOW FIRST FLOOR LINE

HOUSE DRAIN 8'-9" BELOW FIRST FLOOR LINE

Fig. 23-1

ISOMETRIC SKETCH OF PIPING
OF FIGURE 23-1

Unit 24 THE WALL-HUNG TOILET

OBJECTIVES

After completing this unit, the student will be able to:

- discuss the rough-in for a wall-hung toilet.
- describe the combined carrier and soil fittings for wall-hung toilets.

WALL-HUNG TOILET WITH FLUSH VALVE

The wall-hung toilet leaves the floor clear for easy cleaning. It is preferred in industrial, commercial, and school buildings. This toilet style requires a strong and rigid support because of the cantilever forces.

A wall-hung toilet is shown in figure 24-1. The flush valve operation is the same as in other toilet styles and must be used with adequately sized pipe.

CARRIER AND SOIL CONNECTION

The wall-hung toilet is supported by the carrier. The carrier must also give a rigid connection to the soil fitting. The manufacturer recommends that a 1/16-inch clearance be allowed between the face of the finished wall and the back of the bowl. The plumbing rough-in can be done before the wall is built.

Figure 24-2 shows a soil fitting-carrier that can connect into the waste, soil, and vent stack. Figure 24-3 shows a horizontal run from the stack fitting to another wall-hung toilet.

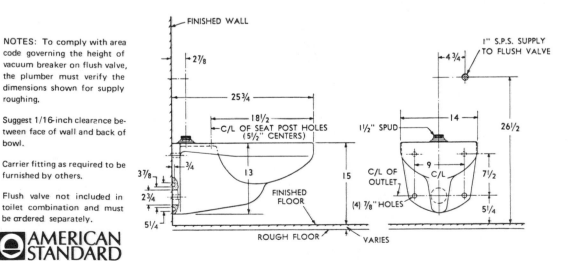

NOTES: To comply with area code governing the height of vacuum breaker on flush valve, the plumber must verify the dimensions shown for supply roughing.

Suggest 1/16-inch clearance between face of wall and back of bowl.

Carrier fitting as required to be furnished by others.

Flush valve not included in toilet combination and must be ordered separately.

AMERICAN STANDARD
PLUMBING & HEATING DIVISION

Fig. 24-1 Vitreous china toilet combination

| | | | | | | | | | L | | | | | |
| DIMENSIONS SUBJECT TO MANUFACTURING TOLERANCE | | | | | | | | | | | | | | |
Product No.	Description	A	B	C	D	E	F	H	Y = 4 ½	Y = 5 ¼	Y = 5 ¾	P*	X‡	Y**
ZU-1204-ISHL4	4" Left Hand ≠≠	4	4	$9\frac{3}{4}$	5	$6\frac{5}{8}$	$9\frac{1}{8}$	$7\frac{1}{2}$	$1\frac{5}{8}$-$3\frac{3}{8}$	$1\frac{5}{8}$-$4\frac{1}{8}$	$1\frac{5}{8}$-$4\frac{5}{8}$			
ZU-1204-ISHR4	4" Right Hand	4	4	$9\frac{3}{4}$	5	$6\frac{5}{8}$	$9\frac{1}{8}$	$7\frac{1}{2}$	$1\frac{5}{8}$-$3\frac{3}{8}$	$1\frac{5}{8}$-$4\frac{1}{8}$	$1\frac{5}{8}$-$4\frac{5}{8}$			
ZU-1204-ISHL54	5" Left Hand	5	4	$10\frac{1}{8}$	$5\frac{1}{2}$	$6\frac{5}{8}$	$9\frac{1}{8}$	$7\frac{3}{4}$	$1\frac{5}{8}$-$3\frac{3}{8}$	$1\frac{5}{8}$-$4\frac{1}{8}$	$1\frac{5}{8}$-$4\frac{5}{8}$	Min.	$4\frac{1}{2}$	
ZU-1204-ISHR54	5" Right Hand	5	4	$10\frac{1}{8}$	$5\frac{1}{2}$	$6\frac{5}{8}$	$9\frac{1}{8}$	$7\frac{3}{4}$	$1\frac{5}{8}$-$3\frac{3}{8}$	$1\frac{5}{8}$-$4\frac{1}{8}$	$1\frac{5}{8}$-$4\frac{5}{8}$	$2\frac{1}{2}$	Min.	$4\frac{1}{2}$
ZU-1204-ISHL55	5" Left Hand	5	5	$10\frac{1}{8}$	$5\frac{1}{2}$	$6\frac{5}{8}$	$9\frac{1}{8}$	$7\frac{3}{4}$	$1\frac{5}{8}$-$3\frac{3}{8}$	$1\frac{5}{8}$-$4\frac{1}{8}$	$1\frac{5}{8}$-$4\frac{5}{8}$			
ZU-1204-ISHR55	5" Right Hand	5	5	$10\frac{1}{8}$	$5\frac{1}{2}$	$6\frac{5}{8}$	$9\frac{1}{8}$	$7\frac{3}{4}$	$1\frac{5}{8}$-$3\frac{3}{8}$	$1\frac{5}{8}$-$4\frac{1}{8}$	$1\frac{5}{8}$-$4\frac{5}{8}$			$5\frac{1}{4}$
ZU-1204-ISHL64	6" Left Hand	6	4	$11\frac{1}{8}$	6	$7\frac{1}{8}$	$9\frac{5}{8}$	$8\frac{1}{4}$	$1\frac{5}{8}$-$3\frac{3}{8}$	$1\frac{5}{8}$-$4\frac{1}{8}$	$1\frac{5}{8}$-$4\frac{5}{8}$			
ZU-1204-ISHR64	6" Right Hand	6	4	$11\frac{1}{8}$	6	$7\frac{1}{8}$	$9\frac{5}{8}$	$8\frac{1}{4}$	$1\frac{5}{8}$-$3\frac{3}{8}$	$1\frac{5}{8}$-$4\frac{1}{8}$	$1\frac{5}{8}$-$4\frac{5}{8}$	See	6"	
ZU-1204-ISHL65	6" Left Hand	6	5	$11\frac{1}{8}$	6	$7\frac{1}{8}$	$9\frac{5}{8}$	$8\frac{1}{4}$	$1\frac{5}{8}$-$3\frac{3}{8}$	$1\frac{5}{8}$-$4\frac{1}{8}$	$1\frac{5}{8}$-$4\frac{5}{8}$	Note	Max.	$5\frac{3}{4}$
ZU-1204-ISHR65	6" Right Hand	6	5	$11\frac{1}{8}$	6	$7\frac{1}{8}$	$9\frac{5}{8}$	$8\frac{1}{4}$	$1\frac{5}{8}$-$3\frac{3}{8}$	$1\frac{5}{8}$-$4\frac{1}{8}$	$1\frac{5}{8}$-$4\frac{5}{8}$			

NOTES: Right Hand Shown.

≠ ≠ For Left Hand, Vent and Hub Inlet Located on Left-Hand Side

≠ 1" Added Adjustment Attained by Rotating Feet.

See Other Side for Optional Faceplate and Coupling Variations.

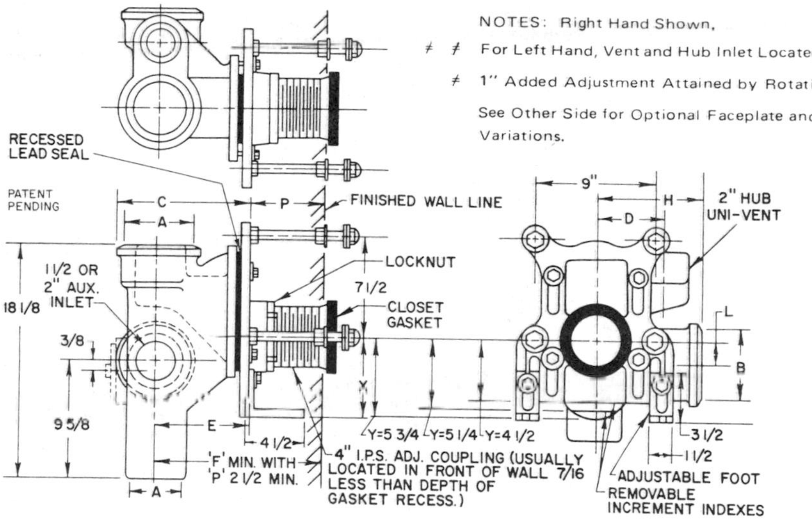

RECESSED LEAD SEAL

PATENT PENDING

18 1/8

1 1/2 OR 2" AUX. INLET

3/8

9 5/8

C
A
P — FINISHED WALL LINE

LOCKNUT
7 1/2
CLOSET GASKET

E
4 1/2
'F' MIN. WITH 'P' 2 1/2 MIN.

Y=5 3/4 Y=5 1/4 Y=4 1/2
4" I.P.S. ADJ. COUPLING (USUALLY LOCATED IN FRONT OF WALL 7/16 LESS THAN DEPTH OF GASKET RECESS.)

A

9"
H
D
2" HUB UNI-VENT

L
B

3 1/2
1 1/2
ADJUSTABLE FOOT
REMOVABLE INCREMENT INDEXES

REGULARLY FURNISHED

Heavy Cast Iron Fitting, Faceplate, Foot Supports, Coupling, Locknut, Mounting Trim

NOTE:

* Regularly Furnished Coupling 'P' Dim. 4 3/4 Min. - 6 1/4 Max. Specify 'P' Dim. for Longer or Shorter Assemblies.

** Indexed Removable Increment Type Faceplate Provides 'Y' Dimension as Required to Meet Specific Closet Roughing Height and Floor Finish.

FURNISHED WHEN SPECIFIED

☐ Longer or Shorter Coupling and Fixture Bolts
☐ Tubular Rigidity Members
☐ 3 1/2 I.P.S. Back Cleanout
☐ Sanitary Aux. Inlet   ☐ V.P. Trim

Zurn Industries, Incorporated

Fig. 24-2 Carrier

DIMENSIONS SUBJECT TO MANUFACTURING TOLERANCE										
Product No.	Description	A-I.P.S.	K	L			N	P*	X ≠	Y**
				Y = 4 1/2	Y = 5 1/4	Y = 5 3/4				
ZEU-1203 SHL4	4"Left Hand	4	$6\frac{1}{8}$	0 - $1\frac{3}{4}$	0 - $2\frac{1}{2}$	0 - 3	$9\frac{3}{4}$	Min. $2\frac{1}{2}$ See Note	$4\frac{1}{2}$ Min. 6" Max.	$4\frac{1}{2}$
ZEU-1203 SHR4	4"Right Hand	4	$6\frac{1}{8}$	0 - $1\frac{3}{4}$	0 - $2\frac{1}{2}$	0 - 3	$9\frac{3}{4}$			$5\frac{1}{4}$
ZEU-1203 SHL5	5"Left Hand	5	$6\frac{5}{8}$	0 - $1\frac{3}{4}$	0 - $2\frac{1}{2}$	0 - 3	$10\frac{1}{8}$			$5\frac{3}{4}$
ZEU-1203 SHR5	5"Right Hand·	5	$6\frac{5}{8}$	0 - $1\frac{3}{4}$	0 - $2\frac{1}{2}$	0 - 3	$10\frac{1}{8}$			

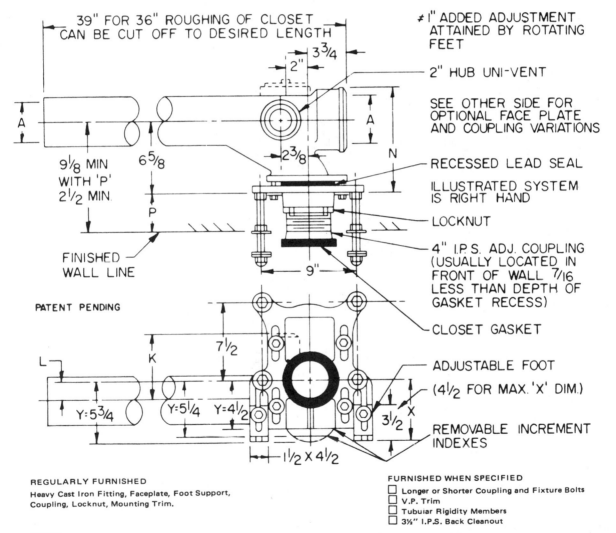

REGULARLY FURNISHED
Heavy Cast Iron Fitting, Faceplate, Foot Support,
Coupling, Locknut, Mounting Trim.

FURNISHED WHEN SPECIFIED
☐ Longer or Shorter Coupling and Fixture Bolts
☐ V.P. Trim
☐ Tubuiar Rigidity Members
☐ 3½" I.P.S. Back Cleanout

NOTE:
* Regularly Furnished Coupling 'P' Dim. 4¾ Min. 6¼ Max. Specify 'P' Dim. for Longer or Shorter Assemblies.
** Indexed Removable Increment Type Faceplate Provides 'Y' Dimension as Required to Meet Specific Closet Roughing Height and
  Floor Finish.

Zurn Industries, Incorporated

Fig. 24-3 Carrier

ASSIGNMENT

Fill-Ins

The following statements refer to figures 24-1, 24-2, and 24-3. Fill in the blanks with the correct information to complete each statement.

1. The rim of the toilet bowl is _____ inches above the floor.

2. The centerline of the waste in figure 24-1 is _____ inches above the floor.

3. Figures 24-2 and 24-3 show that the centerline of the waste can be a minimum of _____ inches and a maximum of _____ inches from the floor.

4. The flush valve requires a _____ -inch standard pipe size supply.

5. The vertical spacing of closet support bolts is _____ inches.

6. The _____ supports the weight of the carrier.

7. A _____ seal is used between the carrier and the fitting.

8. A _____ seal is used between the 4-inch IPS coupling and the toilet bowl.

9. The toilet bowl has a _____ -inch deep gasket recess.

10. The 4-inch IPS coupling will be adjusted to _____ inches in front of the wall.

Unit 25 MEN'S TOILET ROOM—SLAB ON GRADE

OBJECTIVES

After completing this unit, the student will be able to:

- plan pipe assembly for wall-hung toilets.
- plan the soil, waste and vent pipes in a wall space.
- plan plumbing for a slab-on-grade construction.

SLAB ON GRADE

Some buildings are constructed without a basement so that the first floor is concrete poured on a prepared and leveled ground surface. The plumbing which runs under the floor must be installed before the concrete is poured. In this type of construction, a pipe space is included as part of the architect's design.

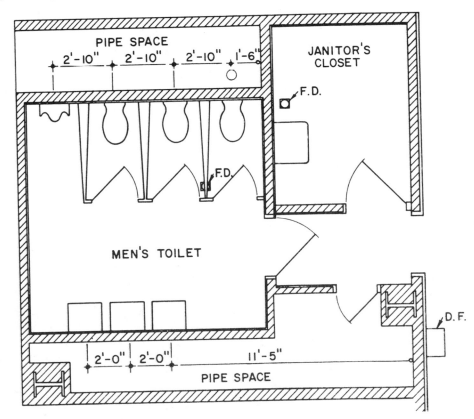

Fig. 25-1

PIPE SIZE

The stack for figure 25-1 is 4 inches, and the house drain is also a 4-inch size. Both the urinal and the service sink use 2-inch wastes. The same pipe sizes are used for the soil, waste, and vent pipes as are used in other construction.

ASSIGNMENT

Drawing Exercises

1. Make an isometric pipe drawing to a convenient scale for the men's toilet room and janitor's closet shown in the plan view of figure 25-1. Rough-in sheets for the urinal, service sink, and drinking fountain are included in the appendix. Show the soil, waste, and vent pipes.

2. Make an isometric water pipe drawing for the same fixtures of figure 25-1.

3. Make an order list of the Zurn Monolithic fittings needed for this toilet room. Refer to the charts in figures 24-2 and 24-3.

SECTION 3
Interpreting Residential Blueprints

Unit 26 VISUALIZING THE HOUSE

OBJECTIVE

After completing this unit, the student will be able to:

- form an overall picture of the house from the architect's plans.

USING RESIDENTIAL BLUEPRINTS

This is the first of nine units in which the plans for a residence are used. They present the blueprint reading and sketching problems that are met by the workers who install the plumbing system. The plans include, in this case, the seven sheets of drawings in Packet 1.

The first step is to study these plans until a reasonable picture of the house is formed. It should be possible to make a complete model of the house and its setting, but usually the model exists only in the mind of the one who reads the working drawings.

ASSIGNMENT

Fill-Ins

The following statements refer to the working drawings in Packet 1. Fill-in the blanks with the correct information to complete each statement.

1. The size of the lot is _____ feet by _____ feet inside the mere (boundary) stones.

2. The greatest lot elevation is _____ feet.

3. The lowest lot elevation is _____ feet.

4. The first floor elevation is _____ feet.

5. The basement floor elevation is _____ feet.

6. The house is set back _____ feet from the front boundary line.

7. The house is located _____ feet from the N-W boundary line.

8. The house measures _____ feet across the front.

9. From the front, the house appears to be _____ stories.

10. From the rear, the house appears to be _____ stories.

11. The basement partitions are made of _____.

12. The hot water tank is located in the _____ room.

13. The gutters and leaders are made of _____.

14. The roofing material shown is _____.

15. The ceilings are shown with _____ -inch thick insulation.

16. The floor joists are _____ by _____ size, spaced _____ on centers.

17. The first floor bath has _____ lavatory basin(s).

18. The front cellar wall is _____ -inch thick concrete.

19. The artesian well water line enters the pump room through _____ -inch thick concrete.

20. The basement floor is _____ inches above the bottom of the footings.

Unit 27 SEWAGE DISPOSAL SYSTEM

OBJECTIVES

After completing this unit, the student will be able to:

- determine elevations for the sewage disposal system in relation to the house.
- estimate quantities of tile pipe for the sewage disposal system.

ELEVATIONS

Elevations are very important to the plumber. Sometimes all excavations for the cellar and for the disposal system are excavated at one time. The finished job must provide proper slopes away from the house.

The engineer measures elevations in feet as 100.25 feet. The plumber may use this system, or feet and inches, as 100'-3". Accuracy is to the nearest .01 foot or nearest 1/8 inch.

SPECIFICATIONS

The high level of the finished grade around the house is to be at elevation 100.00 feet corresponding to elevations on the plot plan of Sheet #1.

The invert of the house drain is to be 7 inches under the house footing at the point where it passes under the footing.

The sewer line to the septic tank shall be 4-inch sewer tile laid at a grade of 1/8 inch per foot sloping to the septic tank. The invert of the sewer tile is to be at least 1'-6" below grade at the inlet to the septic tank.

The septic tank shall have a minimum capacity of 750 gallons and shall be constructed of precast concrete or equal as approved.

The sewer line from the septic tank to the distribution box shall be 4-inch vitrified tile pipe laid at a grade of 1/8 inch per foot.

The distribution box shall be a precast concrete unit or equal as approved.

The seepage bed shall be open joint tile with bends constructed of proper fittings and laid at a grade of 1 inch in 10 feet on a gravel bed of 6-inch minimum thickness. Other pipe materials and perforated type pipe may be substituted for tile as approved.

ASSIGNMENT

Questions

Study the residential blueprints in Packet 1 and answer the following questions.

1. What are the following elevations?

    a. Finished first floor _____

    b. Garage floor _____

    c. Cellar floor   _____

    d. Top of footing   _____

    e. Bottom of footing   _____

    f. Invert of the house drain
       under the footing   _____

2. What is the sewer length from the house to the septic tank?

                                              _____

3. How much drop will this sewer have?   _____

4. What is the elevation of the invert of the sewer at the inlet to the septic tank?

                                              _____

5. What is the length of outflow line from the septic tank to the distribution box?

                                              _____

6. How much drop is there on the outflow sewer line between the septic tank and the distribution box?

                                              _____

7. What is the elevation at the invert of the sewer into the distribution box?

                                              _____

8. How deep in the ground is the invert of the sewer at the inlet to the distribution box?

                                              _____

9. Approximately how many feet of tile are needed for the seepage bed?

                                              _____

Unit 28 SANITARY AND STORM HOUSE SEWERS

OBJECTIVE

After completing this unit, the student will be able to:

- plan the installation of storm and sanitary house sewers.

SANITARY AND STORM HOUSE SEWERS

The storm sewer carries water from rain leaders and other sources that collect rain or run-off water. The storm sewer discharge does not go through a sewage treatment plant.

The sanitary sewer carries waste from each plumbing fixture.

The house sewers must be installed to avoid interferences with water and gas main and the other main sewer.

The plot plan allows for an alternative to septic tank disposal and connections to street sewers. Sometimes health authorities permit septic tank use until street sewers are available.

SPECIFICATIONS

The sanitary sewer shall be of premium joint tile, or its equivalent, and installed at a uniform grade.

The storm sewer shall be of premium joint tile, or its equivalent, and installed at a uniform grade except for the rain leaders which shall be connected to 3-inch cast iron risers. The cast iron shall continue to the tile connections at least 5 feet from the outside wall.

With the architect's permission, rain leaders may be relocated from the southwest side of the house to connect to the storm sewer on the higher level of finished grade.

The garage floor drain shall be connected to the storm sewer.

ASSIGNMENT

Fill-Ins

Refer to the residential blueprints in Packet 1. Make the following calculations and fill in the correct answers.

1. The depth of manhole no. 1 shown on the plot plan is _____.

2. The invert elevation of the street sewer at the point of entry of the house branch is about _____.

3. The invert elevation of the house branch where the fitting connects to the street sewer is about _____.

4. The invert elevation of the house sanitary sewer where it passes under the street water main is about _____.

5. The invert elevation of manhole no. 2 is about _____.

6. The invert elevation of the storm sewer at the point of entry of the house storm sewer is about _____.

7. The invert elevation of the house storm sewer where it connects to a fitting into the street storm sewer is about _____.

8. The invert elevation of manhole A is about _____.

9. The invert elevation of the house storm sewer at the water main is about _____.

10. The invert elevation of the connection to conductor at the northwest corner of the house is about _____.

11. The invert elevation of the connection to the conductor at the southeast corner of the house is about _____.

Drawing Exercise

Make a cross-section drawing of the earth and basement along the line of the house sanitary sewer. Use a horizontal scale of 1″ = 10′-0″ and a vertical scale of 1/4″ = 1′-0″, or scales as assigned by the instructor. Show the correct location of the following:

a. The gas main

b. The water main

c. The street sanitary sewer

d. The street storm sewer

e. The house footing and basement wall

f. The invert of the house sanitary sewer

g. The invert of the house storm sewer

Unit 29 THE HOUSE DRAIN

OBJECTIVE

After completing this unit, the student will be able to:

- plan the house drain with the aid of a mechanical plan.

HOUSE DRAINS

The house drain is considered as a unit in planning the plumbing. In this residence it is installed before the cellar floor is poured.

It is possible to make an isometric pipe drawing of the house drain. It has little rise, however, and is more easily drawn as a two-dimensional drawing. The house drain is thus shown as part of the plumbing mechanical plan for the basement.

The only special convention used is to draw the house trap on its side. This shows it better than drawing it in its proper position.

SPECIFICATIONS

The house drain may be designed for the septic tank system as shown. It may also be designed to connect to a street sewer by using the alternate as shown on the plot plan.

The house drain shall be constructed of 4-inch cast iron pipe and fittings or equal as approved.

Where permissible by code, branches of the house drain may be of 3-inch, service weight cast iron pipe and fittings or equal as approved.

The house trap shall be installed inside the house and have a proper fresh air vent. There shall be suitable, accessible cleanout openings for the house trap. This does not apply where a septic tank is used.

ASSIGNMENT

Drawing Exercises

Study the residential blueprints in Packet 1 and draw the house drain portion of the plumbing mechanical plan for the basement according to the following directions.

1. Make a copy of the basement plan for the residence to the same or different scale. Show partitions, fireplace, and other details that might influence the location of the piping.

2. Draw a satisfactory house drain (single line drawing) to the sanitary sewer on the basement plan. The commonly used symbol is a heavy line with hubs, —3———————3— . An arrow shows direction of flow, but is not needed when the hubs are drawn to show flow direction. In this exercise, the septic tank is not used for disposal.

3. Keep this plan as it will be used again for water piping (Unit 32).

Unit 30 THE WASTE STACK AND VENT PIPING

OBJECTIVE

After completing this unit, the student will be able to:

- draw an isometric pipe drawing for the waste stack for the kitchen sink and laundry trays.

WASTE STACK AND VENT PIPING

When one fixture is located above another and both waste to the same stack, the lower vent must be kept from flooding with discharge from the upper fixture. This is prevented by extending the lower vent to a point above the higher waste connection before connecting to the stack, figure 30-1.

Most codes will allow this vent to be omitted if the higher waste pipe is made one size larger than normally required. For example, 1 1/2" waste pipe must be changed to a 2-inch size, etc. All fixtures on wet vents shall be on the same floor level.

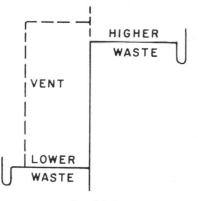

Fig. 30-1

SPECIFICATIONS

The kitchen sink and laundry tray waste system shall conform to local code and provide an adequate, trouble-free waste system that is properly vented.

ASSIGNMENT

Drawing Exercise

Study the blueprints of the residence. Draw an isometric pipe drawing for the waste stack for the kitchen sink and laundry trays.

Unit 31 THE SOIL STACK AND VENT PIPING

OBJECTIVE

After completing this unit, the student will be able to:

- draw an isometric pipe drawing for the soil stack for the two bathrooms of the residence.

GENERAL PROBLEMS

There are three general problems peculiar to bathrooms located one above the other:

1. The partitions are often not directly over each other, therefore requiring an offset in the stack.

2. The vents for the lower fixture must be carried high enough to prevent their flooding by waste from the upper fixtures.

3. Pipe is located in inside partitions so as not to interfere with insulation.

SPECIFICATIONS

The soil stack and branches shall be installed according to local code to provide adequate and durable waste and vent system. The stack shall be the size required by code.

ASSIGNMENT

Drawing Exercise

Study the residential blueprints. Draw an isometric pipe drawing for the soil stack and branches for the bathrooms of the residence.

Unit 32 HOT AND COLD WATER PIPING
IN THE BASEMENT

OBJECTIVE

After completing this unit, the student will be able to:

- plan the hot and cold water piping in the basement.

HOT AND COLD WATER PIPING

The hot and cold water piping is planned for the basement by using the mechanical plan and adding the water lines to it. In order to distinguish one line from another, each has its own symbol as shown in figure 32-1.

SPECIFICATIONS

The water supply can come from a city supply brought into the storage room, as shown on the plans. The city supply is to be metered near the main shut-off in the basement. If a city water supply is used, there shall be a valve installed on each side of the water meter, with a drip on the house side.

All hot and cold water is to be type L copper with sweated fittings. The pipe is to be properly supported to prevent sagging and vibration. All pipe must be of adequate size for the service.

Compression stop and waste valves shall be installed so that each bath, the kitchen sink, or the laundry trays can be controlled for repair without shutting off other parts of the system.

Three hose bibs shall be installed. Each hose bib is to have a stop and waste valve located in the basement:

- Inside the garage
- At the front of the house near the front walk.
- At the rear of the house, approximately at the center of the total length.

Hot and cold water piping shall be run as directly as possible. It shall be supplied with the proper grade and drip opening to permit draining.

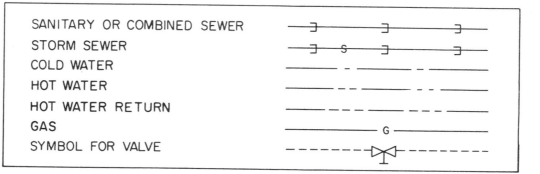

Fig. 32-1

ASSIGNMENT

Drawing Exercise

Add the hot and cold water lines to the mechanical plan for the basement prepared in Unit 29.

Unit 33 WATER PIPING TO THE KITCHEN AND LAUNDRY AND TO THE BATHROOMS

OBJECTIVE

After completing this unit, the student will be able to:

- design a satisfactory water piping system to the kitchen sink and laundry trays and to the bathrooms.

TRADE PRACTICE

The main runs of hot and cold water lines can be shown on a mechanical plan, but the branches to the fixtures are best shown by an isometric pipe drawing. It is possible to add this drawing to the one for waste and vent, but this leads to confusion.

It is good trade practice to make a separate drawing for each major system of the water piping. The water line pipe drawings start at the connection to the main run in the basement and continue to the outlet for the fixture.

ASSIGNMENT

Drawing Exercises

Study the residential blueprints.

1. Make an isometric pipe drawing for hot and cold water to the kitchen sink and laundry trays. Use a suitable scale and show the pipe size.

2. Make an isometric pipe drawing for hot and cold water to the bathroom fixtures on the first floor. Use a suitable scale and show the pipe size.

3. Make an isometric pipe drawing for hot and cold water to the bathroom fixtures in the basement. Use a suitable scale and show the pipe size.

Unit 34 GAS PIPING

OBJECTIVE

After completing this unit, the student will be able to:

- design a satisfactory gas piping system for the residence.

Gas piping is often installed in houses where gas is available even though it may not be planned to use gas at the outlets provided at the time of construction.

SPECIFICATIONS

Gas pipe is to be installed to provide for the following outlets:

- Hot water heater
- Heating boiler or furnace
- Kitchen stove
- Log lighter to each fireplace

The gas piping shall be black steel pipe with threaded fittings installed in a quality manner and tested for leakage. The following sizes are required:

- One-inch main from the meter in the storage room to the boiler

- A 3/4-inch branch to the kitchen stove

- The other outlets may terminate in 1/2-inch pipe with branches from the main at 3/4-inch size for more than one outlet or for runs of over 10 feet.

ASSIGNMENT

Drawing Exercises

Study the residential blueprints.

1. Add the gas piping to the basement mechanical plan using approved line symbols.

2. Draw an isometric pipe drawing of the entire gas piping in the residence using a proper scale and showing the pipe size.

SECTION 4
Commercial Building Blueprints

Unit 35 PLOT PLANS AND THE ENGINEER'S SCALE

OBJECTIVE

After completing this unit, the student will be able to:

- use the civil engineer's scale.

THE COMMERCIAL BUILDING

This is the first of several units based on the working drawings in Packet 2 and the specifications in the appendix for a commercial building. The steps outline what a plumber needs to do in order to plan the work. The ideas, requirements of plumbing design, and procedures are much the same as those already covered. There are new problems, however, and different factors to consider in planning the commercial job. It is much cheaper to change a line on paper than to move a waste stack or other plumbing.

Most commercial buildings are larger than the one shown. The size makes it possible to use 1/4" = 1'-0" scale without the need for larger sheets of paper. The architect presents commercial construction on sheets usable for classroom study.

THE ENGINEERING RULE
OR TAPE

The practice in land measure is to use a steel tape which has each foot divided into ten parts (.1 foot) and each tenth divided into ten more parts (.01 foot). Length measure by feet and decimal feet is practical for civil engineers and surveyors. Most plot plans are so dimensioned.

THE ENGINEER'S SCALE

Plot plans are often drawn with a scale that differs from that used by the architect. The engineer's scale has sides with scales of 10, 20, 30, 40, and 50 parts to an inch. Thus, a detail plot plan might be drawn to a scale of 1" = 10'-0". A very large plot plan might be drawn to a scale of 1" = 50'-0".

It is not too difficult to use an ordinary rule to draw plot plans. At a scale of 1" = 10'-0":

$$1/2'' = 5' \text{ or } 60''$$
$$1/4'' = 2 \ 1/2' \text{ or } 30''$$
$$1/8'' = 1 \ 1/4' \text{ or } 15''$$
$$1/16'' = 5/8' \text{ or } 7 \ 1/2''$$

Thus, 68 feet is shown as a line 6 3/4 inches long. By following the above illustration, a relationship for each of the engineer's scales can be established.

ASSIGNMENT

Reading Plot Plans

1. Study the plot plan for the commercial building found in Packet 2. Note that the architect has located the building using feet and inch measure.

2. Determine from the plot plan and specifications, the location of sewer, water, and gas services to the building.

3. Compare the plot plans for the commercial building and the residence. Note the use of contour lines.

4. Draw the plot plan to a scale of 1″ = 10′-0″. (Use scale or rule). Locate sewer, water and gas mains, and each service to the building.

Unit 36 AN OVERALL PICTURE
OF THE BUILDING

OBJECTIVES

After completing this unit, the student will be able to:

- form an overall picture of the building from the architect's plans and specifications.
- discuss the need for planning ahead of construction.

THE COMMERCIAL INSTALLATION

Whether a building is residential, commercial, or industrial can change some details of construction. However, the plumbing waste and vent systems follow the same design patterns in all buildings. There are often some plumbing code requirements based on the proposed use of the building, but a store and office building is, for the most part, just a larger plumbing installation than a residence.

BUILDING CONSTRUCTION MATERIALS

The plumber must plan somewhat differently for a building according to the construction materials for floors and walls. Wood, concrete or masonry, and steel all present different problems for the plumber.

Wood framing in floors and walls is easily cut to allow passage of pipe through holes or notches. Often the plumber is able to work after the rough carpentry is done. However, advance planning permits small changes to avoid excessive cutting and weakening of the structure.

Concrete and masonry can be cut, but it is a more expensive process than cutting wood. It is easier and cheaper to install sleeves, chases, and inserts during construc-

tion. This requires that the plumbing work be planned before construction begins.

Sleeves are short pieces of pipe placed in concrete or masonry walls to provide an opening for piping. Where several pipes close together pass through a concrete partition or floor, a square or rectangular opening can be left by having the carpenter install a box in the form. The sleeve or boxed opening serves two purposes. First, it avoids expensive cutting during construction; and, second, it allows for pipe expansion and replacement.

Piping is sometimes installed before masonry partitions are in place. In this case, a sleeve should be placed on each pipe so that it will be built into the partition or wall by the bricklayers. The outside wall can be made watertight between sleeve and pipe by caulking the space at both ends of the sleeve.

A *chase* is a space to conceal risers in a masonry wall. The chase may be a recess so that the finished wall is straight, or the finished wall may have to be thicker to enclose the piping. It is the plumber's responsibility in many cases to locate the chase and determine its size for the necessary pipe risers and stacks, figures 36-1 and 36-2.

Steel beams and columns should not be cut to aid the plumbing installation. Plumbing

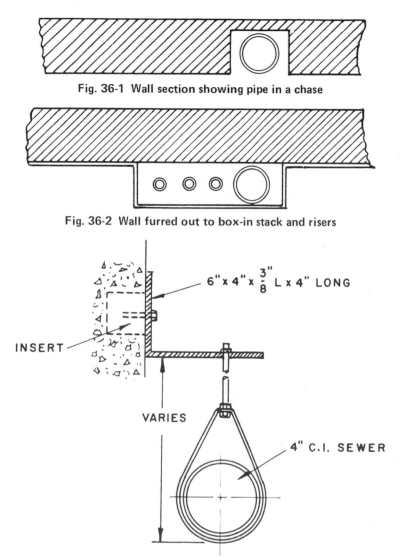

Fig. 36-1 Wall section showing pipe in a chase

Fig. 36-2 Wall furred out to box-in stack and risers

$6" \times 4" \times \frac{3}{8}"$ L x 4" LONG

INSERT

VARIES

4" C.I. SEWER

Fig. 36-3 Pipe hanger detail, wall mounting

design must avoid them if at all possible. The reinforcement steel in concrete is much less of a problem to the plumber who installs sleeves during construction rather than attempting to cut an opening after the concrete has been poured.

Pipe is supported in many installations on hangers or brackets. The inserts to attach hangers or brackets to concrete or masonry can be installed by drilling holes. There are also inserts that are located in the form before the concrete is poured, figures 36-3 and 36-4.

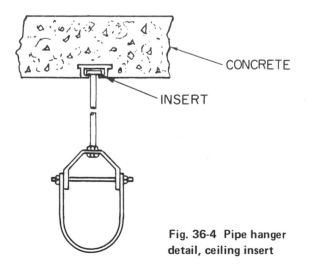

CONCRETE

INSERT

Fig. 36-4 Pipe hanger detail, ceiling insert

ASSIGNMENT

Questions

Answer the following questions based on a study of the commercial building plans and specifications in Packet 2 and the appendix.

1. What is the length and width of the building? _____

2. How many stories are there in the building? _____

3. What use is the basement designed for? _____

_____

4. What use is the first floor designed for? _____

5. What use is the second floor designed for? _____

6. How will the ground level for pouring the basement floor be established?

_____

7. How many basement floor drains are specified? _____

8. What do the floor drains connect into? _____

9. Is there reinforcement steel in the:

    a. front basement wall?   _____
    b. side walls of the basement?   _____
    c. rear basement wall?   _____
    d. basement partition?   _____

10. How many crosswise steel beams support the first floor? _____

11. What lengthwise steel beams are used to support the first floor? _____

_____

12. What is the material and the size of framing of the first floor other than the steel beams? _____

13. How many windows does the basement have? _____

14. How many rain leaders or conductors are specified? _____

15. In your locality would the rain leaders or conductors connect to: (circle answer)

    a. House drain
    b. Storm drain
    c. Dry well, discharge to ground

16. What stores are located on the first floor? _____

17. What are the plumbing fixtures for each store?

_____

_____

_____

18. What is the construction material for the side and rear walls of the store? _____

19. What is the construction material for the dividing partition? _____ _____

20. How are the risers to the second floor to be concealed in the dividing partition? _____

21. Which businesses have offices on the second floor? _____ _____

22. What are the plumbing fixtures for the second floor? _____ _____

23. What is the construction material for the second floor partitions? _____

24. What size space will there be between floor joists for pipe runs to the second floor plumbing? _____

25. How are the second floor lavatories lighted and ventilated? _____ _____

26. What is the attic height over the lavatories? _____

27. What roofing material is to be used? _____

28. What type of material will be used for roof flanges? _____

29. List the passages for pipe through floors, walls, and partitions shown by the architect. _____

30. List any additional plumbing that is necessary or desirable. _____ _____

Unit 37 ELEVATIONS AND THE PLUMBING INSTALLATION

OBJECTIVE

After completing this unit, the student will be able to:

- solve elevation problems in relation to the plumbing for the commercial building.

SEPARATE OR COMBINED DRAINS

In planning elevations for a house drain, it is necessary to determine if storm and sanitary drainage are separate or combined. In some parts of the country, building codes require separate systems.

When allowed, a combined sewer in the street takes off both storm and sanitary flow. Thus, all stacks, floor drains, and conductors connect into one house drain. A sanitary sewer or a septic tank disposal requires that sanitary and storm flow have separate piping.

For many buildings, there is a storm or clear water drain to the nearest stream, figure 37-1. The elevations are then specified by the architect and are to be established and maintained during construction.

ELEVATION OF STREET SEWER

The elevation of the main sewer in the street in relation to the elevation of the basement floor is one of the first considerations in designing the house drain.

A low street sewer permits the house drain to be under the basement floor.

A high sewer requires that basement floor drains flow to a sump. A sump pump lifts the water to the house drain. If it is nesessary to discharge a water closet to a sump, a sewage ejector should be used.

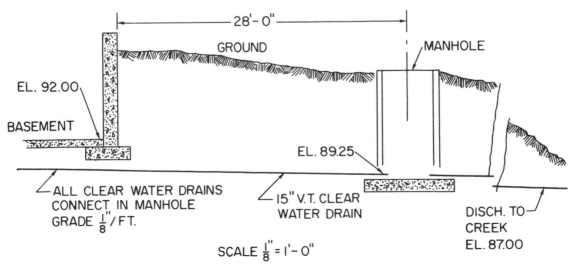

Fig. 37-1 Clear water (storm) drain

INVERT OF SEWER

Most pipe lines are measured to a center-line of the pipe. Elevations of sewers are measured to the invert of the sewer, which is the inside bottom of the pipe.

SAMPLE PROBLEM

Basement finished floor elevation   200.00'
First floor elevation   209.15'
Main sewer at elevation invert   203.87'

Thus: 203.87'
              −200.00'

Invert of sewer is   3.87' above basement floor.

.87 × 12 = 10.44" or invert of sewer is 3'-10 1/2" above the cellar floor as measured on a rule.

GRADE AND ELEVATION

A sewer or house drain should not be installed level but at a slight downgrade.

Often a grade of 1/8 inch per foot is specified. This is almost the same as a grade of .01 inch per foot as 1/8 inch is nearly equal to .01 foot.

A house sewer 40 feet long would change in elevation as follows:

40 × 1/8" = 5" or 40 × .01' = .40'

CONNECTION OF HOUSE SEWER TO MAIN SEWER

There are various fittings and connectors for joining the house sewer to the main sewer. The elevation change at the connection depends somewhat on the depth of the main sewer.

The minimum elevation difference, for most situations, is equal to the difference in diameters of the sewer. This puts the tops of the two sewers at the same level. Thus, when a 6-inch sewer connects to a 10-inch main sewer, the invert of the house sewer is 4 inches or .33 foot above the invert of the main sewer.

ASSIGNMENT

Drawing Exercises

1. Determine the following elevations from the working drawings and specifications for the commercial building. Make calculations as needed.

    a. Elevation of street sewer   _____
    b. Elevation of house sewer at invert
       through front basement wall   _____
    c. Elevation of basement floor   _____
    d. Elevation of first floor   _____

2. Make a drawing of the house sewer using a horizontal scale of 1/4" = 1'-0" and a vertical scale of 3/4" = 1'-0". Label the elevations.

3. Make an elevation view of the inside of the front basement wall showing dimensioned locations for the sleeves through which the house drain, water supply, and gas supply pass.

For class use	North from outside South wall	Sewer 9'-0" Water 16'-9" Gas 12'-6"	Elevation	Sewer code or higher Water 4' cover or code Gas 2' cover or code

Unit 38 THE HOUSE DRAIN

OBJECTIVE

After completing this unit, the student will be able to:

- design a house drain for the commercial building.

HOUSE DRAIN DESIGN

Study of the commercial building should show the following about the design of the house drain:

- The house drain is above the basement floor.
- Exposed piping would be permissible in the basement used for storage.
- The location and number of fixtures are shown.
- The problems of concealing stacks and risers.
- Whether to combine or keep separate the sanitary and storm flow.

In order to design the drain, these additional factors must be considered:

- The drain must be of adequate size to meet code and trade standards.
- The location must be such that runs are as direct as possible.
- The location must be such that piping does not interfere with use of space.
- The location must be planned to avoid pipe in front of doors or windows.
- Changes in elevation of the house drain may be required due to grade.
- Elevations of the branches must give head room clearance.
- Cleanouts must be designed to facilitate maintenance.

ASSIGNMENT

Drawing Exercises

Study the working drawings of the commercial building.

1. Make a copy of the basement plan to the same or different scale. Start a mechanical plan by showing a line representation of the house drain. Label openings for stacks as stack #1, etc. Note elevations at points for easy location on the job.

2. Show details by separate drawings or by an isometric drawing of the house drain.

Unit 39 SOIL STACK, WASTE AND VENT PIPING

OBJECTIVES

After completing this unit, the student will be able to:

- plan the waste and vent stacks and branches.
- detail the chases needed to conceal the stacks and risers.

BASIC INFORMATION

The design of waste and vents for the commercial building is similar to that of the residence. It is also similar to the problems in the isometric drawing section of this book. The materials must conform to the specifications and to the plumbing code that applies to the locality.

The new problems in this building are how to avoid steel beams and conceal the piping. The concrete and block construction leaves little pipe space.

ASSIGNMENT

Drawing Exercises

1. Draw isometric drawings of soil and waste stacks as required for the plumbing of the commercial building. Note pipe sizes and materials on the drawings.

2. When not shown on plans, make detail drawings of each chase required showing size of the chase and its location.

3. Make detail drawings of any sleeve locations needed for installation of the waste and vent stacks.

Unit 40 HOT AND COLD WATER PIPING

OBJECTIVE

After completing this unit, the student will be able to:

- plan the water piping for the commercial building.

WATER PIPE SIZE

In a commercial building with several tenants, the water supply must be large enough to maintain an adequate flow of water to all fixtures at all times. Each fixture manufacturer shows on the rough-in sheet the size of pipe needed to adequately supply that fixture. For instance, a water closet flush valve must have a 1-inch supply, whereas a flush tank closet needs only a 1/2-inch supply. (See the rough-in sheets in the appendix for commercial building fixtures.)

Because of friction in long lengths of pipe and the friction caused by fittings, pipe lines are often larger than the final connection to the fixture.

The architect must judge the use factor of plumbing in a building as all outlets are expected to run full flow at the same time. Refer to article No. 25 in the specifications for commercial plumbing and gas installation for a chart of pipe sizes as recommended by the architect. Follow the recommendations as shown for minimum pipe sizes. Larger sizes are recommended for the water piping if greater flow is required.

The water pipe materials are indicated in the specifications unless a written agreement for a change is made.

ASSIGNMENT

Drawing Exercises

Study the commercial building working drawings.

1. Draw the water piping in the basement on the mechanical plan for the basement. Label riser openings as first floor sink, second floor toilet rooms, etc.

2. Draw isometric drawings of the water piping from basement outlets to fixtures. Do not try to show so much piping in one drawing that it becomes cluttered. Make several drawings as necessary.

3. Use notes so that the relationship of one drawing to another is clearly understood.

Unit 41 GAS PIPING

OBJECTIVE

After completing this unit, the student will be able to:

- plan the gas piping for the commercial building.

GAS PIPING

Gas piping is simpler than other piping. It is best to plan its installation after the water lines and waste systems are located. It is comparatively easy to alter the gas piping to avoid the other systems.

Gas lines are usually black iron pipe, although plastic can sometimes be used underground. Gas piping should be of adequate size and should run as directly as possible. The flow of gas, like the flow of liquids, is retarded by friction, turns, and restrictions.

In most towns and cities the gas is owned and operated by a utility company. The utility company installs the gas service and the meter in most cases. The utility company can also provide technical information as to gas pipe size or other problems that confront the plumber.

It is the responsibility of the plumbing contractor to pay the utility company for installing the gas service and meter in the building.

ASSIGNMENT

Drawing Exercises

Study the commercial building working drawings and specifications.

1. Study the plans and specifications for points of gas use. Make a list of gas outlets that are needed.

2. Use the basement mechanical plan and add the gas piping runs in the basement to this plan. Label each outlet as water heater, etc.

Unit 42 CONTINUED PRACTICE IN BLUEPRINT READING

OBJECTIVE

After completing this unit, the student will be able to:

- start a program for the continued study of blueprint reading.

Completing the assignments in this text is just the start of the study of blueprint reading. Blueprint reading must be practiced to reinforce the skill. It is also a never-ending learning process because building designs and architectural methods are always being revised and updated.

ARCHITECTURAL METHODS

The architect's office, like any business, is on the lookout for timesaving methods. Thus the symbols in use today are different from those previously used and may change again. (See the Architectural Symbols in the appendix.)

For example, the water closet symbol is now two rectangles. It once was a rectangle and a circle and even a rectangle and an oval. A continued study of plans will keep the plumber informed of variations as they occur.

PLUMBING IN INSTITUTIONS

Many buildings have plumbing problems not included in the residence or the commer-

cial building. Large plumbing installations, cramped space, and plumbing specialties require that the architect provide more than the customary information to the plumber. The architect will often provide additional drawings to help solve such problems.

There are four general practices used to show more plumbing information on plans:

1. Plumbing fixtures may be outlined in both plan and elevation views. This establishes the locations of fixtures, especially where space is at a minimum.

2. The architect may draw a suggested mechanical plan to show the layout in a new building or, in the case of remodeling or additions to a building, to show existing piping.

3. In some cases an isometric pipe drawing is included. This is to help the plumber, but is not an absolute specification.

4. Special details from manufacturers, such as swimming pool water treatment installations, are included in some plans.

ASSIGNMENT

Questions

A. Study the partial plans for a hospital, figures 42-1 and 42-2, pages 113 and 114, and answer the following questions.

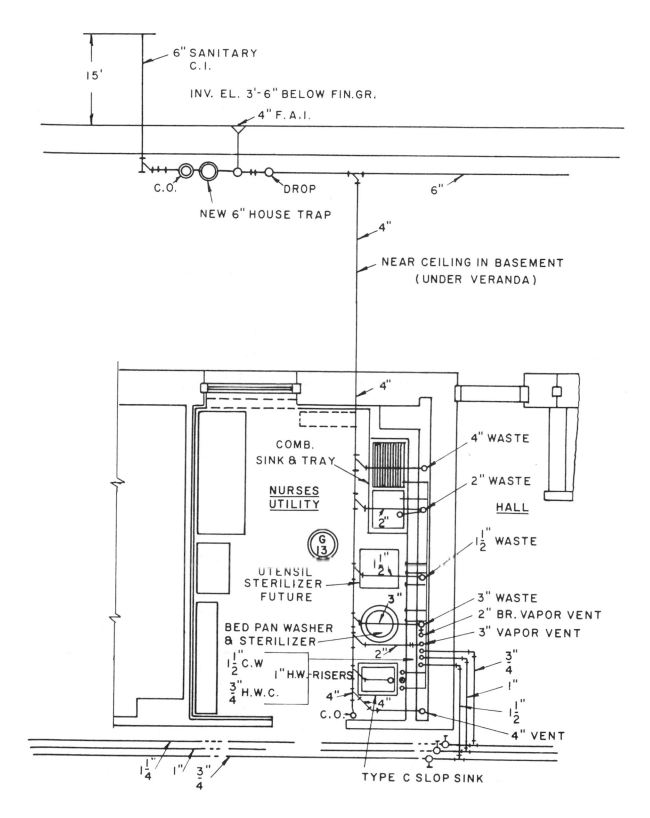

Fig. 42-1

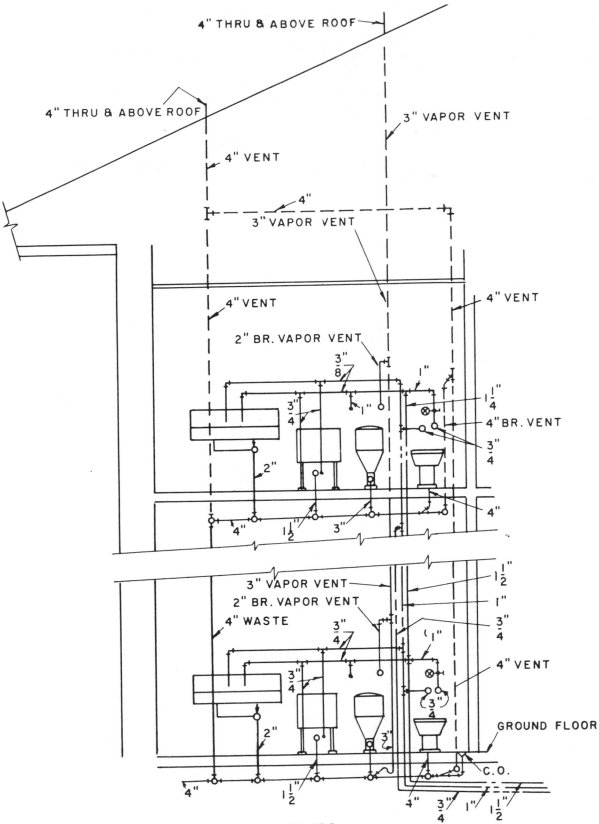

Fig. 42-2

1. List the plumbing fixtures shown on one floor.

    _____

    _____

    _____

    _____

    _____

    _____

    _____

    _____

    _____

2. How has the architect helped the plumber?

    _____

3. What size vent stack is shown? _____

4. What is a vapor vent and what fixtures does it serve?

    _____

    _____

5. How is the vapor vent installed in relation to the other plumbing?

    _____

    _____

B. Study additional plans and specifications as provided by the instructor. Look for differences in drafting techniques and for plumbing information.

Appendix A

SYMBOLS

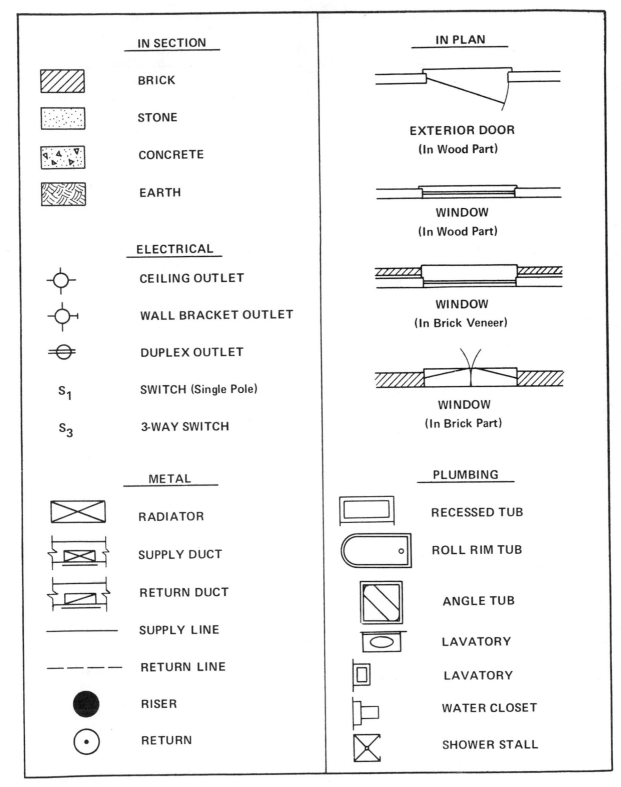

IN SECTION

BRICK

STONE

CONCRETE

EARTH

ELECTRICAL

CEILING OUTLET

WALL BRACKET OUTLET

DUPLEX OUTLET

S_1 SWITCH (Single Pole)

S_3 3-WAY SWITCH

METAL

RADIATOR

SUPPLY DUCT

RETURN DUCT

SUPPLY LINE

RETURN LINE

RISER

RETURN

IN PLAN

EXTERIOR DOOR
(In Wood Part)

WINDOW
(In Wood Part)

WINDOW
(In Brick Veneer)

WINDOW
(In Brick Part)

PLUMBING

RECESSED TUB

ROLL RIM TUB

ANGLE TUB

LAVATORY

LAVATORY

WATER CLOSET

SHOWER STALL

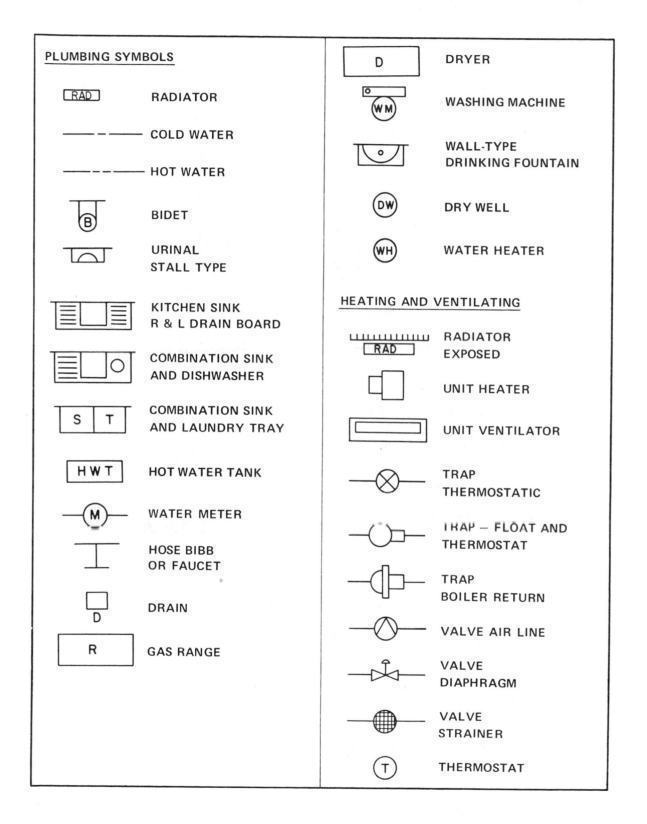

PLUMBING SYMBOLS

RAD — RADIATOR

—— – —— COLD WATER

—— – – —— HOT WATER

BIDET

URINAL
STALL TYPE

KITCHEN SINK
R & L DRAIN BOARD

COMBINATION SINK
AND DISHWASHER

S | T — COMBINATION SINK
AND LAUNDRY TRAY

H W T — HOT WATER TANK

M — WATER METER

HOSE BIBB
OR FAUCET

DRAIN
D

R — GAS RANGE

D — DRYER

WM — WASHING MACHINE

WALL-TYPE
DRINKING FOUNTAIN

DW — DRY WELL

WH — WATER HEATER

HEATING AND VENTILATING

RAD — RADIATOR
EXPOSED

UNIT HEATER

UNIT VENTILATOR

TRAP
THERMOSTATIC

TRAP – FLOAT AND
THERMOSTAT

TRAP
BOILER RETURN

VALVE AIR LINE

VALVE
DIAPHRAGM

VALVE
STRAINER

T — THERMOSTAT

Appendix B
BOCA Plumbing Code*

SECTION P-1205.0 WATER CLOSETS

P-1205.1 Approval: All water closets shall be of the water conservation type.

P-1205.2 Water closets for public use: Water closet bowls for public use shall be of the elongated type.

P-1205.3 Water closet seats: Water closets shall be equipped with seats of smooth, nonabsorbent materials. All seats for water closets provided for public or employee use shall be of the hinged open-front style. Integral water closet seats shall be of the same material as the fixture. Water closet seats shall be sized for the water closet bowl type.

P-1205.4 Water closet connection: A 4-inch by 3-inch closet bend shall be acceptable. Where a 3-inch bend is used on a water closet, a 4-inch by 3-inch flange shall be used to receive the fixture horn.

SECTION P-116.0 WORKMANSHIP

P-116.1 General: All work shall be conducted, installed and completed in a workmanlike and acceptable manner so as to secure the results intended by this code and the standards referenced herein.

SECTION P-602.0 DRAINAGE PIPING INSTALLATION

P-602.1 Pitch of horizontal drainage piping: Horizontal drainage piping shall be installed in uniform alignment at uniform slopes. The minimum pitch of a horizontal drainage pipe shall be in accordance with Table P-602.1

Table P-602.1
PITCH OF HORIZONTAL DRAINAGE PIPES

Size (inches)	Minimum pitch (inches per foot)
1½ or less	1/2
2 to 2½	1/4
3 to 6	1/8
8 or larger	1/16

Note: 1 inch per foot = 0.0833 mm/m.

SECTION P-905.0 VENT GRADES AND CONNECTIONS

P-905.1 Vent grade: All vent and branch vent pipes shall be graded and connected so as to drain back to the soil or waste pipe by gravity.

DRAINAGE PIPE CLEANOUTS

SECTION P-1100.0 GENERAL

P-1100.1 Scope: The provisions of this article shall control the size and location of drainage pipe cleanouts, and provide requirements for their installation and maintenance.

SECTION P-1101.0 WHERE REQUIRED

P-1101.1 Horizontal drains within buildings: All horizontal drains 4 inches in diameter or less shall have cleanouts located not more than 50 feet (15240 mm) apart. All horizontal drains larger than 4 inches in diameter shall have cleanouts located not more than 100 feet (30480 mm) apart.

P-1101.2 Building sewers: All building sewers shall have cleanouts located not more than 100 feet (30480 mm) apart.

P-1101.3 Changing direction: Accessible cleanouts shall be installed at each change of direction of the building drain, or of horizontal waste and vent lines, which is greater than 45 degrees (0.79 rad.).

P-1101.4 Base of stack: A cleanout shall be provided at the base of each waste or soil stack.

P-1101.5 Building drain and building sewer junctions: There shall be a cleanout at the junction of the building drain and the building sewer. This cleanout shall be either inside or outside the building wall, and shall be brought up to finish grade or to the basement floor level. The cleanout at the junction of the building drain and the building sewer shall not be required if the cleanout on a 3-inch or larger diameter vertical soil stack is located within 10 feet (3048 mm) of the building drain and building sewer connection.

P-1001.7 Building traps: Building (house) traps shall not be installed except where specifically required by the plumbing official. Each building trap, when installed, shall be provided with a cleanout and with a relieving vent or fresh air intake on the inlet side of the trap which need not be larger than one half the diameter of the drain to which it connects. Such relieving vent or fresh air intake shall be carried above grade and terminate in a screened outlet located outside the building.

SANITARY DRAINAGE SYSTEMS

P-601.1 Drainage Load: The drainage fixture unit load shall be determined for each sanitary drainage pipe in accordance with Table P-601.1a, Table 601.1b, and Section P-601.1.1. Fixtures not listed in Table P-601.1a shall have a drainage fixture unit load based on the outlet size of the fixture in accordance with Table P-601.1b. The minimum trap size for unlisted fixtures shall be the size of the drainage outlet, but not less than 1 1/4 inches.

Table 601.1a
DRAINAGE FIXTURE UNIT VALUES FOR VARIOUS PLUMBING FIXTURES

Type of fixture or group of fixtures	Drainage fixture unit values (dfu)	Trap size in inches
Automatic clothes washer standpipe	3	2
Bathroom group	6	
Bathtub	2	1½
Bidet	1	1¼
Combination sink and tray	2	1½
Dental unit	1	1¼
Dishwasher	2	1½
Drinking fountain	½	1¼
Floor drains	2	2
Kitchen sink	2	1½
Laundry tray	2	1½
Lavatory	1	1¼
Shower (each head)	2	2
Sink	2	1½
Urinal	4	2
Water closet, nonpublic	4	—
Water closet, public	6	—
Water closet pneumatic assist private or public installation	4	—

Table P-601.1b
DRAINAGE FIXTURE UNIT VALUES FOR FIXTURE DRAINS OR TRAPS

Size (inches)	Drainage fixture unit value
1¼ or less	1
1½	2
2	3
2½	4
3	5
4	6

P-601.1.1 Values for continuous flow: Drainage fixture unit values for continuous or semicontinous flow into a drainage system, such as a pump, sump ejector, air conditioning equipment, or similar device, shall be computed on the basis of one fixture unit equalling 7½ gallons per minute (28.35 lpm).

VENTS AND VENTING

SECTION P-901.0 WHERE REQUIRED

P-901.1 Individual vent: Each trap and trapped fixture shall be provided with an individual vent, except as otherwise approved by this code.

P-901.2 Trap seal: The vent system shall be designed and installed so that the trap seal shall be subject to a maximum pneumatic pressure differential equal to 1 inch (25 mm) water column.

SECTION P-902.0 VENT PIPE SIZES

P-902.1 General: Vent stacks and stack vents shall be sized in accordance with Table P-902.1 on the basis of length and drainage load computed from Table P-601.1a, Table P-601.1b, and Section P-601.1.1.

P-902.2 Branch vents: Branch vents shall be sized in accordance with Table P-902.2 based on the vent pipe developed length and the diameter of the horizontal drainage pipe.

P-902.3 Individual vents: The diameter of the individual vent shall be at least one-half the diameter of the drain served, except a vent pipe shall not be less than 1¼ inches in diameter.

P-902.4 Relief vents: The diameter of a relief vent shall be at least one-half the diameter of the soil waste branch served.

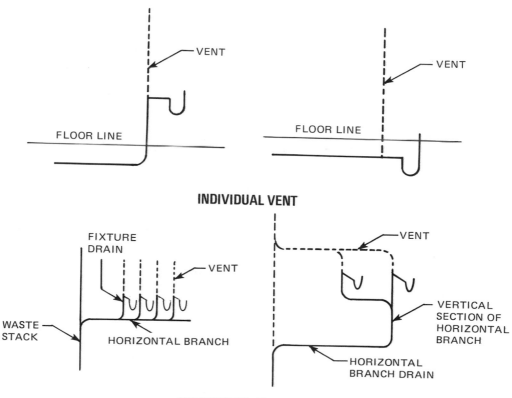

INDIVIDUAL VENT

HORIZONTAL BRANCH

Table P-902.1
SIZE AND LENGTH OF VENT STACKS AND STACK VENTS

Diameter of soil or waste stack (in.)	Total fixture units connected to stack (dfu)	Diameter of vent (inches)[a] (Maximum developed length of vent, in feet, given below)										
		1¼	1½	2	2½	3	4	5	6	8	10	12
1¼	2	30										
1½	8	50	150									
1½	10	30	100									
2	12	30	75	200								
2	20	26	50	150								
2½	42		30	100	300							
3	10		42	150	360	1040						
3	21		32	110	270	810						
3	53		27	94	230	680						
3	102		25	86	210	620						
4	43			35	85	250	980					
4	140			27	65	200	750					
4	320			23	55	170	640					
4	540			21	50	150	580					
5	190				28	82	320	990				
5	490				21	63	250	760				
5	940				18	53	210	670				
5	1,400				16	49	190	590				
6	500					33	130	400	1000			
6	1,100					26	100	310	780			
6	2,000					22	84	260	660			
6	2,900					20	77	240	600			
8	1,800						31	95	240	940		
8	3,400						24	73	190	720		
8	5,600						20	62	160	610		
8	7,600						18	56	140	560		
10	4,000							31	78	310	960	
10	7,200							24	60	240	740	
10	11,000							20	51	200	630	
10	15,000							18	46	180	570	
12	7,300								31	120	380	940
12	13,000								24	94	300	720
12	20,000								20	79	250	610
12	26,000								18	72	230	500
15	15,000									40	130	310
15	25,000									31	96	240
15	38,000									26	81	200
15	50,000									24	74	180

Note a. To convert feet to millimeters, multiply by 304.8.

Table P-902.2
MINIMUM DIAMETERS AND MAXIMUM LENGTH OF INDIVIDUAL, BRANCH, CIRCUIT, AND LOOP VENTS FOR HORIZONTAL SOIL AND WASTE BRANCHES

Diameter of horizontal branch (inches)	Slope or horizontal branch[a] (inches per foot)	Diameter of vent (inches) (Maximum developed length of vent, in feet, given below)[a]									
		1¼	1½	2	2½	3	4	5	6	8	10
1¼	1/8	NL[b]									
	1/4	NL									
	1/2	NL									
1½	1/8	NL	NL								
	1/4	NL	NL								
	1/2	NL	NL								
2	1/8	NL	NL	NL							
	1/4	290	NL	NL							
	1/2	150	380	NL							
2½	1/8	180	450	NL	NL						
	1/4	96	240	NL	NL						
	1/2	49	130	NL	NL						
3	1/8		190	NL	NL	NL					
	1/4		97	420	NL	NL					
	1/2		50	220	NL	NL					
4	1/8			190	NL	NL	NL				
	1/4			98	310	NL	NL				
	1/2			48	160	410	NL				
5	1/8				190	490	NL	NL			
	1/4				97	250	NL	NL			
	1/2				46	130	NL	NL			
6	1/8					190	NL	NL	NL		
	1/4					96	440	NL	NL		
	1/2					44	220	NL	NL		
8	1/8						190	NL	NL	NL	
	1/4						91	310	NL	NL	
	1/2						38	150	410	NL	
10	1/8							190	500	NL	NL
	1/4							85	240	NL	NL
	1/2							32	110	NL	NL
12	1/8								180	NL	NL
	1/4								79	420	NL
	1/2								26	200	NL

Note a. To convert feet to millimeters, multiply by 304.8; to convert inches per foot to centimeters per meter, multiply by 8.33.
Note b. NL means No Limit. Actual values in excess of 500 feet.

P-903.1 Vent stack required: Every building in which plumbing is installed shall have at least one main stack, not less than 3 inches in diameter, which shall run undiminished in size and as directly as possible from the building drain through to the open air above the roof. A vent stack or a main vent shall be installed with a soil or waste stack whenever individual vents, relief vents, or other branch vents are required in a building of two or more branch intervals.

SECTION P-906.0 WET VENTING SYSTEMS

P-906.1 Single bathroom groups: A single bathroom group of fixtures is permitted to be intalled with the drain from an individually vented lavatory or combination fixture serving as a wet vent for the bathtub or shower compartment and for a water closet. Wet vent systems shall conform to Sections P-906.1.1 and P-906.1.2.

P-906.1.1 Wet vent size: A 1 1/2-inch diameter wet vent shall drain a maximum of one fixture unit. A 2-inch diameter wet vent shall drain a maximum of four fixture units.

P-906.1.2 Wet vent connection: A wet vent drainage pipe shall connect to the upper half of a horizontal soil pipe or water closet bend at a maximum angle of 45 degrees (0.79 rad.) from the direction of flow. When connection is to a stack, the wet vent drainage pipe shall connect at the same level as the water closet or below the water closet.

P-906.2 Double bathroom groups, back to back: Where bathroom groups back to back, consisting of two lavatories and two bathtubs or shower combinations, are installed on the same horizontal branch with a common vent for the lavatories and without vent for the bathtubs or shower compartments and the water closets, the wet vent shall not be less than 2 inches in diameter, and the length of the fixture drain shall conform to Table P-909.1.

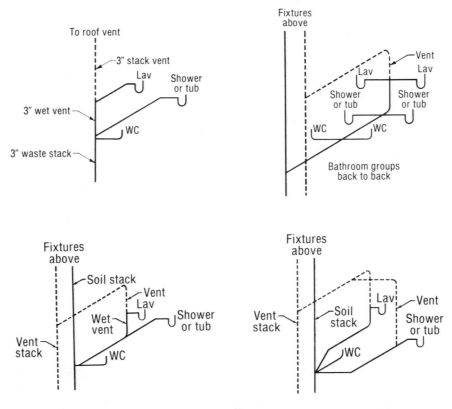

WET VENT

SECTION P-909.0 FIXTURE VENTS

P-909.1 Distance of trap from vent: Each fixture trap shall have a protecting vent so located that the developed length of the fixture drain from the trap weir to the vent fitting is within the requirements set forth in Table P-909.1.

Table P-909.1
MAXIMUM DISTANCE OF FIXTURE TRAP FROM VENT[a]

Size of trap	Size of fixture drain	Fall per foot	Distance from trap
1¼″	1¼″	1/4″	3′ 6″
1¼″	1½″	1/4″	5′
1½″	1½″	1/4″	5′
1½″	2 ″	1/4″	8′
2 ″	2 ″	1/4″	6′
3 ″	3 ″	1/8″	10′
4 ″	4 ″	1/8″	12′

Note a. To convert inches per foot to centimeters per meter, multiply by 8.33; to convert feet to millimeters, multiply by 304.8.

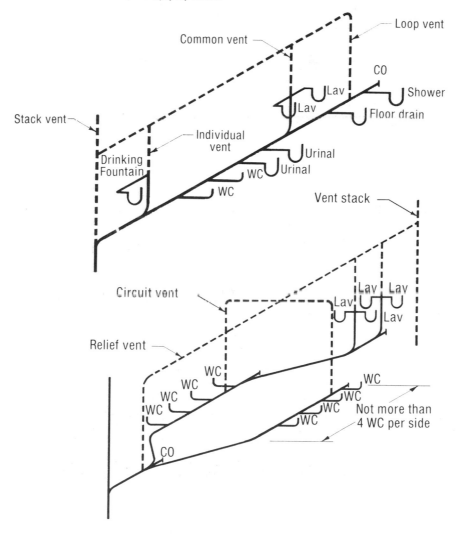

CIRCUIT AND LOOP VENTING

STORM DRAINAGE SYSTEMS

SECTION P-801.0 WHERE REQUIRED

P-801.1 General: All roofs paved areas, yards, courts, and courtyards shall be drained into a separate storm sewer system, or a combined sewer system where such systems are available, or to an approved place of disposal. In the case of one and two family dwellings, where approved by the plumbing offical, storm water is permitted to be discharged onto flat areas such as streets or lawns provided the storm water will flow away from the building.

P-801.2 Cleanouts required: Cleanouts shall be installed in the storm drainage system and shall comply with the provisions of this code on drainage pipe cleanouts.

SECTION P-802.0 STORM WATER DRAINAGE TO SANITARY SEWER PROHIBITED

P-802.1 General: Storm water shall not be drained into sanitary sewers.

SECTION P-803.0 SIZE OF BUILDING STORM DRAINS, BUILDING STORM SEWERS AND LEADERS

P-803.1 Size of horizontal drains and sewers: The size of the building storm drains or building storm sewers or any of their horizontal branches shall be based upon the maximum projected roof or paved area to be drained. The size shall be determined in accordance with Table P-803.1.

Table P-803.1
SIZE OF HORIZONTAL BUILDING STORM DRAINS AND BUILDING STORM SEWERS[a]

Diameter of drain inches	Maximum projected area in square feet and[b] gallons per minute flow for various slopes					
	1/8 in. per ft. slope		1/4 in. per ft. slope		1/2 in. per ft. slope	
	Square feet[a]	gpm	Square feet[a]	gpm	Square feet[a]	gpm
3	3288	34	4640	48	6576	68
4	7520	78	10600	110	15040	156
5	13360	139	18880	196	26720	278
6	21400	222	30200	314	42800	445
8	46000	478	65200	677	92000	956
10	82800	860	116800	1214	165600	1721
12	133200	1384	188000	1953	266400	2768
15	238000	2473	336000	3491	476000	4946

Note a. Table P-803.1 is based upon a maximum rate of rainfall of 1 inch (25 mm) per hour for a one hour duration and a 100 year return period. The figure for drainage area shall be adjusted to local conditions by dividing by the local rate in inches per hour. See Appendix F.

Note b. To convert inches per foot to centimeters per meter, multiply by 8.33; to convert square feet to square meters, multiply by 0.093; to convert gallons per minute to liters per minute, multiply by 3.78.

P-111.5.1 Type of plans: Plans and specifications, where required by the code officials, shall include plan view and a riser diagram showing the work. Such plans shall show the direction of flow, pipe size, grade of horizontal piping, elevations, draining fixture unit loads of both stacks and drains in the drain waste and vent systems, and the supply fixture unit load for the water system and any branch supplies which serve more than one plumbing fixture, appliance or hose outlet.

P-111.6 Site plan: There also shall be filed a site plan showing the location of water service and sewer connection with respect to any building in which a plumbing system is to be installed. Vent stack termination shall be shown with respect to building ventilation openings which could allow introduction of sewage gasses into the building or any adjacent building.

P-111.6.1 Private sewage disposal systems: The site plan shall indicate the location of a private sewage disposal system when a public sewer is not available. The site plan shall include all technical data and soil data required by the private sewage disposal code——.

P-115.1 Required inspections: The code official shall conduct inspections during and upon completion of the work for which a permit has been issued. A record of all such examinations and inspections and of all violations of this code shall be maintained by the code official.

P-115.2 Final inspections: Upon completion of the plumbing work, a final inspection shall be made. All violations of the approved plans and permits shall be noted, and the holder of the permit shall be notified of the discrepancies.

P-115.1.2.3 Test and inspection records: All required test and inspection records shall be accessible to the code official at all times during the fabrication of the plumbing system and the erection of the building, or such records as the code official designates shall be filed.

P-119.1 Approval: After the prescribed tests and final inspection indicate the work complies in all respects with the code, a notice of approval shall be issued by the code official.

Appendix C

SPECIFICATIONS — PLUMBING AND GAS INSTALLATION (COMMERCIAL)

1. GENERAL

The contractor for plumbing is expected to visit the site and be familiar with existing conditions. Either that part of the plans or the specifications whichever is the greater shall be part of the contract.

2. COOPERATION WITH OTHER CONTRACTORS

The plumbing contractor shall perform his work to conform to construction called for under other contracts and he shall install and relate his work at all times and in such a manner as to cooperate with the others and not to delay or interfere with the work of any other contractor. Any material stored in the building must be distributed to prevent overloading of the floors or blocking of passageways and as directed by the Architect.

3. SCOPE OF WORK

The work included in the Contract is shown on the plans or described in these specifications and shall consist of furnishing all labor and material for the construction, erection, installation, connection and completion of sewers and drainage systems for sanitary and storm sewers.

A 2″ water service shall be extended to the water main in the street.

A 1 1/2″ gas service shall be extended to the gas main in the street.

Water main is approximately 4′ from the curb.

Gas main is approximately 2′ from the curb.

4. INSPECTION

This contractor must have all work inspected and approved by the proper authorities before it is concealed and when required, submit written proof of such approval to the Architect.

This contractor must lay out all work carefully and accurately and be responsible for the proper working of same. He must protect all his work after placing and make all openings around pipe and other service connections watertight in an approved manner.

5. WORKMANSHIP AND MATERIALS

All materials, fixtures, fittings, appliances and apparatus must be of a late pattern, the best kind specified and free of all defects which would injure their efficiency, durability or appearance.

All work must be performed in a neat, skillful and workmanlike manner in accordance with the best modern practice for each class of work and to the satisfaction of the Architect. The work shall be installed in strict accordance with the Plumbing Code.

6. DRAWINGS AND SPECIFICATIONS

The drawings and specifications together are intended to provide for the installation of a complete, first class plumbing and drainage system, water supply, distribution system and gas supply system.

Structural conditions may require certain modifications in piping and pipe runs. Such modifications are permissible if approved by the Architect, but pipe sizes or necessary specific requirements for the satisfactory working of the systems as designed shall not be changed.

This contractor shall not proceed with any work without sufficient drawings or instructions and in case he has not been supplied with them, he shall apply to the Architect for such drawings and explanations as may be necessary.

The following are the drawings referred to by these specifications and declared to be a part of this contract:

Plot plan	Sheet 1
Basement plan	Sheet 2
1st floor plan	Sheet 3
2nd floor plan	Sheet 4
Elevations and details	Sheets 5-11

7. LAYOUT OF SYSTEM

Before the work is commenced, this contractor shall submit for approval a complete layout of the proposed system of sewers, drains, soil stacks, vents and piping. Where any deviation from the plan is required, indicate clearly the exact position of all chases, openings in foundation wall, trenches for drains, etc., with proposed size of all pipes shown. One copy of this layout shall be filed in Architect's office for approval. After approval, a copy shall be placed in the hands of the General Contractor.

8. PROTECTION OF APPARATUS AND WORK

This contractor shall at all times take such precautions as may be necessary to properly protect all apparatus, fixtures, fittings and installations under this contract from damage of any kind. All openings on piping system shall be securely capped or plugged until final connections are made.

9. EXCAVATIONS

This contractor shall make all excavations necessary to receive his work, including all trenches outside or inside of building below the line of other contractors' excavations.

After the work is installed, inspected and approved by the Architect, all excavations, trenches, etc., shall be refilled and the backfill thoroughly tamped down in layers not to exceed 12" in thickness, after permission has been granted by the Architect.

10. SANITARY AND STORM SEWERS (OUTSIDE)

A 6" PVC schedule 40 sewer shall be connected to the main sewer in Main Street and extended to 5' from the building wall. Pipe and fittings for sanitary sewer shall be PVC pipe, installed complete by plumbing contractor as shown. Excavation, backfilling, and joints between PVC shall be done by the plumbing contractor as herein specified.

Furnish and install as required 1/8 regulation bends, sanitary or curved tees, etc., for all changes in course of sewer, and use Y junctions for all branches.

Pipe shall set firmly on solid ground for the entire length, soil being removed under hubs to accomplish this. The first 8″ of backfill shall be #2 washed stone, or equal.

Any indicated elevations of sewer lines on drawings refer to inside bottom of sewer.

Sewer lines shall be laid in directions indicated on drawings with a uniform pitch of 1/4″ per foot unless otherwise authorized by the Architect.

Solvent cement shall be used in making joints between PVC pipes or pipe and fitting. Manufacturer's instructions and plumbing code requirements are to be used for good workmanship.

11. HOUSE DRAINAGE SYSTEM

All cast iron pipe and fittings shall be tarred, service weight, and of size indicated on drawings.

Any elevation of sewer indicated on drawings refers to inside bottom of pipe.

The cast iron pipe and fittings shall be of hub and spigot type designed for use with an approved compression gasket and assembled using the push seal procedure.

Joints shall be water and gas tight to meet any required tests.

12. CLEANOUTS

Furnish and install cleanouts on all lines of iron pipe sewers where shown, at changes in direction and where otherwise necessary to give complete access to all sewer runs for cleaning pipes of obstructions. Cleanouts shall have regulation bends where necessary with extra heavy brass ferrule with tap screw leaded into same.

13. FLOOR DRAINS (BASEMENT AND AREA)

Furnish and install where shown cast iron drain, bottom calked outlet, heavy duty grate. Zurn A-500.

14. BUILDING DRAINS INSIDE OF BUILDING

All sanitary drains and storm water drains to a point 5′ outside building shall be service weight cast iron pipe.

15. SOIL AND WASTE STACKS

All soil and waste stacks 3″ or over in size shall be service weight cast iron pipe and fittings; waste lines smaller than 3″ shall be galvanized wrought iron pipe and galvanized drainage fittings.

Each stack shall have a TY and cleanout at the foot with extra heavy brass ferrule and trap screw. All horizontal waste pipe runs shall be provided with brass cleanout plug where change of direction occurs.

Stacks shall be properly supported at each floor with wrought iron clamps bolted around pipe and securely fastened.

Where fixture wastes are suspended from ceiling, the branch waste connection to each fixture shall be carried to the ceiling so as to conceal connection in floor construction.

All soil and waste stacks shall be erected plumb and straight with a minimum amount of fittings and with slip seal joints using a proper gasket.

16. VENT STACKS

Furnish and install vent stacks to properly vent all fixtures and equipment of size shown on drawings or herein specified and not less than size required by code.

Vent stacks shall be carried to a height not less than 2' above the highest fixture served and connected into soil stack.

Soil stacks and separate or independent vent stacks shall be continued through roof and flashed with 4# sheet lead dressed into top of pipe and made weatherproof.

Vent pipe is to be the same material as mentioned for soil and waste stacks with the exception that fittings are to be galvanized malleable iron.

17. WROUGHT IRON PIPES

Wrought iron pipe wherever specified shall be Byers or approved equal, standard weight, genuine wrought iron pipe.

Couplings and nipples shall be made from same quality iron as pipe.

18. PIPE HANGERS

Pipes suspended from floors and ceilings shall be hung on approved adjustable ring pipe hangers; pipes on side walls shall be supported by approved pipe brackets.

Hangers and brackets shall be spaced not over 6' apart and pipes shall be grouped to give a neat and orderly appearance.

Hangers suspended from concrete floor or roof construction shall be hung from iron inserts furnished and placed in the forms by the contractor before concrete is poured. If a hanger is required where an insert has not been provided, hanger shall be secured through concrete slab with proper washer and not above slab. Gun type inserts are not acceptable.

Where hangers are suspended from structural steel member, they shall be hooked or slipped to member in a secure manner satisfactory to Architect.

19. LOCATION OF PIPING

Stacks, wastes, vents, etc., in finished parts of building shall be concealed wherever possible.

20. SUMP PUMP

Furnish and install in boilerroom a submersible sump pump. Pump shall have capacity of 40 gpm at 15 ft. head, with water proof cables and plug. Furnish and install a C.I. or steel cover for a 24" V.T. sump.

21. TEST-ROUGHING

The entire system of cast iron sewer, soil waste and vent piping shall be subjected to a water test in the presence of the Architect and subjected to such other tests as may be required.

The entire piping shall be absolutely tight and not show any leaks. No pipe shall be covered until all required tests and inspections have been made and the same approved by the Architect.

The acceptance of the work shall not, however, prejudice any claim which the owner may have for the replacement of defective material or workmanship and in accordance with the provisions of the "Guaranty" under these specifications.

22. COLD WATER MAIN SUPPLY CONNECTIONS

Plumbing contractor shall make all arrangements with the local Water Department to pay all costs to install a 2" K-type copper water service from the main. Place approved gate valve inside building wall and install meter. *Note:* The use of 95-5 solder will avoid lead contamination. This precaution is required by some codes.

23. COLD WATER SUPPLY RISERS

Furnish and install all cold water piping as shown and/or herein specified. Cold water piping to be L-type copper pipe with cast or wrought copper sweat fittings. Make connection with main supply in basement.

24. HOT WATER SUPPLY

Pipe used for hot water lines shall be L-type copper tubing with cast or wrought copper sweat fittings.

Furnish and install all hot water piping as herein specified. Start at the hot water storage tank in boiler room; furnish and install main and connections to water risers.

Connect ends of supply mains to a circulating return and extend the return to boiler room; connect same to hot water storage tank.

Connections as required, for even circulating and uniform supply, shall be made to all fixtures shown or specified. Provide main supply at storage tank with gate drainage valves with 3/4" hose ends for complete drainage of system.

25. COLD AND HOT WATER SUPPLY BRANCHES

Take off branch supplies from cold and hot water supply lines or risers to supply all fixtures in the building as follows:

Fixtures	Cold	Hot
Lavatories	1/2"	1/2"
Hose bibbs	3/4"	
Slop and service sinks	3/4"	3/4"

In determining size of branch mains, risers and main supplies for cold and hot water, contractor shall be governed by this table:

Line to Supply	Number	Branches
1/2''	2	1/2''
3/4''	4	1/2''
3/4''	1	3/4''
	2	1/2''
1''	4	3/4''
1 1/4''	2	1''
1 1/2''	2	1 1/4''

Exception shall by made to the above sizes for lines that supply flush valve for water closets which shall by proportioned as follows:

1 water closet-tank type	2	1/2''
1 water closet-flush valve	1	1 1/4''
2 to 4 closets-flush valve	1	1 1/2'' with 1 1/4'' branch to each

These sizes shall be increased wherever in the judgment of the contractor said increase may be desirable for proper operation of the closet valve. Vertical line supplying batteries shall be increased to one size larger than the above schedule.

All concealed water lines shall be firmly secured.

26. DOMESTIC HOT WATER SYSTEM

Storage tank shall be a horizontal steel tank, 200-gallon capacity, 125# W.P. as manufactured by Cemline Corporation or approved equal. Tank shall have openings for hot, cold water, drain, hot water heater recirculating lines, pressure temperature relief valve: install pressure and temperature relief valve of adequate capacity as specified by valve manufacturer.

Hot water heater shall be 1 Ruud #5.

Electric power will be brought to snap switch on adjacent wall by electrical contractor. Plumbing contractor shall provide and pay to make electric connection to aquastat, motor and heaters.

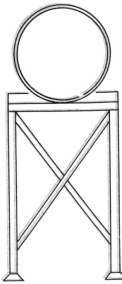

Breeching for hot water heater shall be 20 gauge galvanized steel, all installed by plumbing contractor.

Tank stand to be made of 2'' pipe legs supporting tank saddles. The storage tank to rest on the saddles. See sketch.

27. HOT WATER CIRCULATING PUMP

Furnish and install on hot water circulating lines as indicated in boiler room a 3/4" bronze body Bell and Gossett or approved equal centrifugal booster pump, directly connected to a 1/5 h.p. single-phase, 60-Hz., 110-volt motor. Pump shall be controlled by a Minneapolis-Honeywell L-444A or approved equal immersion-type aquastat. Electric power will be brought to snap switch on adjacent wall by electrical contractor. Plumbing contractor shall provide and make electric connections to aquastat and motor.

28. GAS SUPPLY SYSTEM

The plumbing contractor will make such service agreement with Heat, Light and Power Company as may be required and pay all costs for the installation of a 1 1/2" gas service from the main in the street to the meter in the building.

29. GAS PIPING

Install 1 1/2" gas line from the gas meter and extend same to the hot water heater in the boiler room. Gas pipe to be black steel pipe with black malleable fittings.

Appendix D

FIXTURES FOR THE COMMERCIAL BUILDING

Sink—Drug Store and Bakery
    American Standard 7013.030
    4146.130 Bottom Mount Faucet
    4311.015 waste 1 1/2"

Water Closet—Drug Store and Bakery
    American Standard 2109. Ser. Elongated Cadet Toilet
    With 3/8" angle slip joint stop
    Church Seat #9500 white

Water Closet—2nd floor toilet rooms
    American Standard 2221.018 Madera with Sloan
    #110 Flush Valve
    Church Seat #9500 white

Lavatories—except in 2nd floor men's room
    All lavatories (4) American Standard 0355.012
    Lucerne with 2103 Series Fittings,
    flexible supplies, and 1 1/4" trap

Lavatory—2nd floor men's room
    American Standard 0451.021 Corner Minette Lavatory
    with 2103 Series Fittings,
    3/8" flexible supplies, and 1 1/4" P trap

Slop Sink—2nd floor janitor's room
    American Standard 7696.016 Akron
    with 8344.111 faucet
    and 7798.176 trap standard

Appendix E

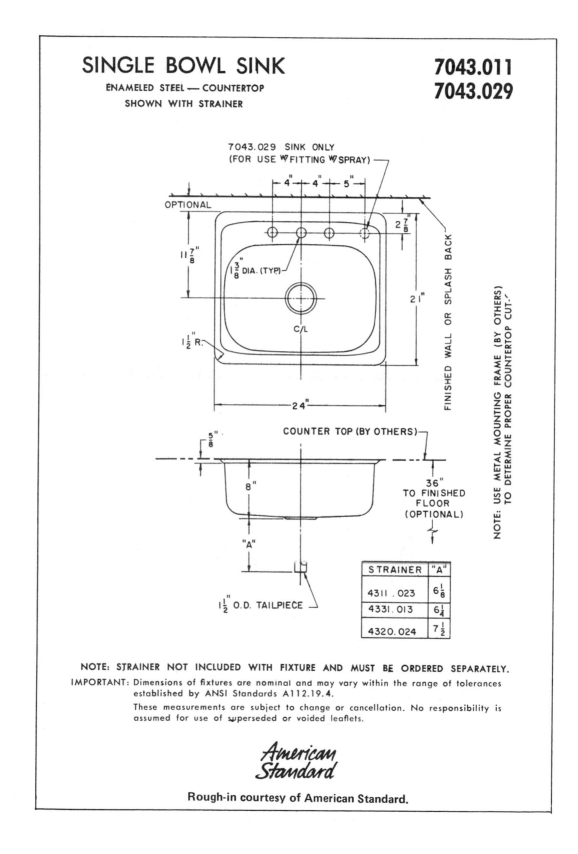

SINGLE BOWL SINK

ENAMELED STEEL — COUNTERTOP
SHOWN WITH STRAINER

7043.011
7043.029

7043.029 SINK ONLY
(FOR USE W/ FITTING W/ SPRAY)

4" — 4" — 5"

OPTIONAL

$2\frac{7}{8}$"

$11\frac{7}{8}$"

$1\frac{3}{8}$" DIA. (TYP)

21"

FINISHED WALL OR SPLASH BACK

$1\frac{1}{2}$" R.

C/L

24"

$\frac{5}{8}$"

COUNTER TOP (BY OTHERS)

8"

36"
TO FINISHED FLOOR
(OPTIONAL)

"A"

$1\frac{1}{2}$" O.D. TAILPIECE

NOTE: USE METAL MOUNTING FRAME (BY OTHERS)
TO DETERMINE PROPER COUNTERTOP CUT.

STRAINER	"A"
4311.023	$6\frac{1}{8}$
4331.013	$6\frac{1}{4}$
4320.024	$7\frac{1}{2}$

NOTE: STRAINER NOT INCLUDED WITH FIXTURE AND MUST BE ORDERED SEPARATELY.

IMPORTANT: Dimensions of fixtures are nominal and may vary within the range of tolerances established by ANSI Standards A112.19.4.

These measurements are subject to change or cancellation. No responsibility is assumed for use of superseded or voided leaflets.

American Standard

Rough-in courtesy of American Standard.

CUSTOM-LINE SINK

7012.024/.032
7013.014/.030

ENAMELED CAST IRON — COUNTERTOP
SHOWN WITH-   RELIANT , TOP MOUNT
OR BOTTOM MOUNT FAUCET &
4331.013 STRAINER

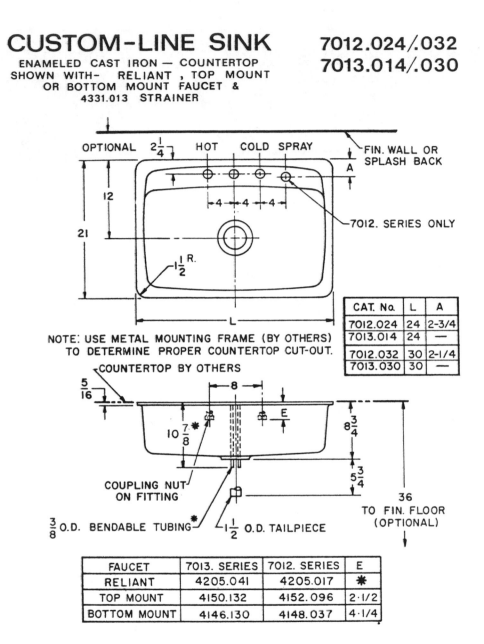

NOTE: USE METAL MOUNTING FRAME (BY OTHERS)
TO DETERMINE PROPER COUNTERTOP CUT-OUT.

CAT. No.	L	A
7012.024	24	2-3/4
7013.014	24	—
7012.032	30	2-1/4
7013.030	30	—

FAUCET	7013. SERIES	7012. SERIES	E
RELIANT	4205.041	4205.017	✱
TOP MOUNT	4150.132	4152.096	2·1/2
BOTTOM MOUNT	4146.130	4148.037	4·1/4

NOTE: FITTINGS NOT INCLUDED WITH FIXTURE AND MUST BE ORDERED SEPARATELY.

IMPORTANT: Dimensions of fixtures are nominal and may vary within the range of
tolerances established by ANSI Standards A112.19.1-M

These measurements are subject to change or cancellation.
No responsibility is assumed for use of superseded of voided leaflets.

American Standard

Rough-in courtesy of American Standard.

ELONGATED CADET TOILET   2109. SER.

VITREOUS CHINA — CLOSE-COUPLED COMBINATION

REGULAR RIM HEIGHT
SHOWN WITH
3/8 FLEX. SUPPLY

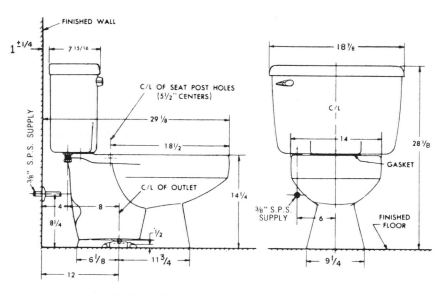

PLUMBER NOTE: THIS COMBINATION IS DESIGNED TO ROUGH IN AT MINIMUM DIMENSION OF 12" FROM FINISHED WALL TO C/L OF OUTLET.

DIMENSIONS SHOWN FOR SUPPLY ARE SUGGESTED

NOTE: 3/8" supply pipe not included with toilet and must be ordered separately.

IMPORTANT: Dimensions of fixtures are nominal and may vary within the range of tolerances established by ANSI Standards A112.19.2. M

These measurements are subject to change or cancellation. No responsibility is assumed for use of superseded or voided leaflets.

Rough-in courtesy of American Standard.

MADERA TOILET
VITREOUS CHINA SHOWN WITH
SLOAN 110 OR DELANY 402 FLUSH VALVE

2221.018
2221.026

NOTE: 2221.026 HAS SLOTS IN RIM FOR BEDPAN HOLDING.

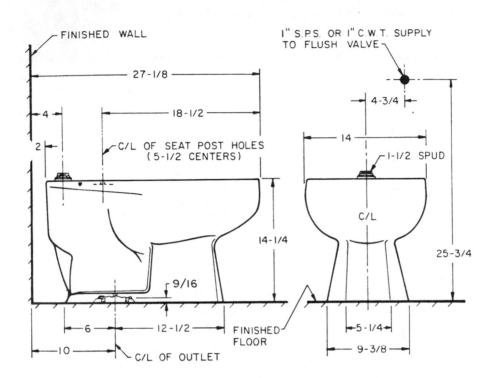

NOTE: TO COMPLY WITH AREA CODE GOVERNING THE
HEIGHT OF VACUUM BREAKER ON FLUSH VALVE,
THE PLUMBER MUST VERIFY DIMENSIONS SHOWN
FOR SUPPLY ROUGHING.

NOTE: FLUSH VALVE NOT INCLUDED WITH TOILET AND MUST BE ORDERED SEPARATELY.

IMPORTANT: Dimensions of fixtures are nominal and may vary within the range of
tolerances established by ANSI Standards A112.19.2. M

These measurements are subject to change or cancellation.
No responsibility is assumed for use of superseded of voided leaflets.

American Standard

Rough-in courtesy of American Standard.

BIDET

SHOWN WITH 1825.012 DUALUX
BIDET FITTING
AND *3/8 FLEX SUPPLIES

5005. SER.

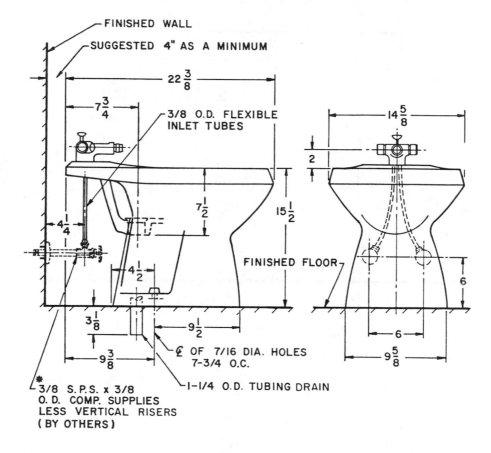

NOTE: SUPPLIES NOT INCLUDED WITH FIXTURE AND MUST BE ORDERED SEPARATELY. (BY OTHERS)

IMPORTANT: Dimensions of fixtures are nominal and may vary within the range of
tolerances established by ANSI Standards A112.19.2 M

These measurements are subject to change or cancellation.
No responsibility is assumed for use of superseded of voided leaflets.

American Standard

Rough-in courtesy of American Standard.

LUCERNE LAVATORY   0355.012

VITREOUS CHINA - FOR CONC. ARM OR WALL
HUNG SUPPORT — SHOWN WITH 2103 OR 2350
SER. FTGS. 3/8 FLEX SUPP. & 1-1/4 x 1-1/4 O.D. "P" TRAP

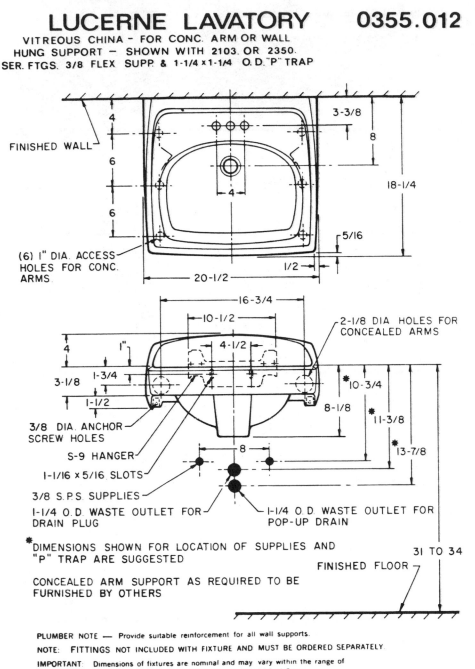

PLUMBER NOTE — Provide suitable reinforcement for all wall supports.

NOTE: FITTINGS NOT INCLUDED WITH FIXTURE AND MUST BE ORDERED SEPARATELY.

IMPORTANT: Dimensions of fixtures are nominal and may vary within the range of
tolerances established by ANSI Standards A112.19.2.M

These measurements are subject to change, or cancellation
No responsibility is assumed for use of superseded of voided leaflets

American
Standard

Rough-in courtesy of American Standard.

OVATION LAVATORY

ENAMELED STEEL — SELF-RIMMING

3004.207
(POP-UP DRAIN)

SHOWN WITH

2103. SERIES ⎤
2385. SERIES ⎦ FITTING   3/8" FLEX. SUPPLIES
1-1/4 X 1-1/4 O.D. "P" TRAP

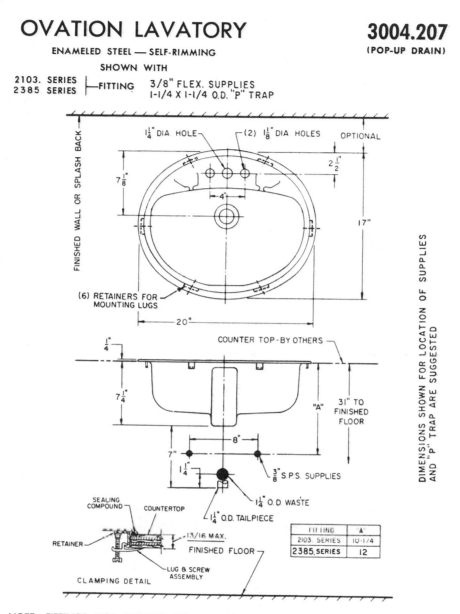

FITTING	"A"
2103. SERIES	10-1/4
2385. SERIES	12

NOTE: FITTINGS NOT INCLUDED WITH FIXTURE AND MUST BE ORDERED SEPARATELY.

NOTE: SEE PAGE 42 FOR COUNTERTOP CUTOUT INSTRUCTIONS

IMPORTANT: Dimensions of fixtures are nominal and may vary within the range of tolerances established by ANSI Standards — A112.19.4.M

These measurements are subject to change or cancellation. No responsibility is assumed for use of superseded or voided leaflets.

American Standard

Rough-in courtesy of American Standard.

CORNER MINETTE LAVATORY   0451.021

VITREOUS CHINA — WALL HUNG   (POP-UP DRAIN)
SHOWN WITH

2103. SERIES ┐
2385. SERIES ┘ ├FITTING

3/8" FLEX. SUPPL
1-1/4 X 1-1/4 O.D. "P" TRAP

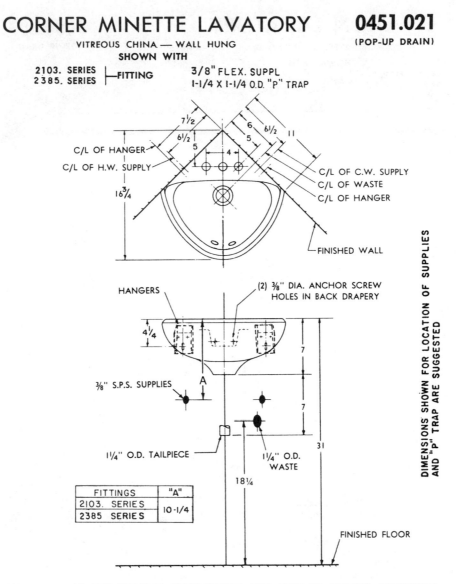

C/L OF HANGER
C/L OF H.W. SUPPLY

C/L OF C.W. SUPPLY
C/L OF WASTE
C/L OF HANGER

FINISHED WALL

HANGERS

(2) 3/8" DIA. ANCHOR SCREW HOLES IN BACK DRAPERY

3/8" S.P.S. SUPPLIES

A

1¼" O.D. TAILPIECE

1¼" O.D. WASTE

FINISHED FLOOR

DIMENSIONS SHOWN FOR LOCATION OF SUPPLIES AND "P" TRAP ARE SUGGESTED

FITTINGS	"A"
2103. SERIES	10-1/4
2385 SERIES	

NOTE: FITTINGS NOT INCLUDED WITH FIXTURE AND MUST BE ORDERED SEPARATELY.

PLUMBER NOTE — Provide suitable reinforcement for all wall supports.

IMPORTANT: Dimensions of fixtures are nominal and may vary within the range of tolerances established by ANSI Standards A112.19.2. M

These measurements are subject to change or cancellation. No responsibility is assumed for use of superseded or voided leaflets.

American Standard

Rough-in courtesy of American Standard.

WHEELCHAIR LAVATORY 9141.011

VITREOUS CHINA — FOR CONC. ARM SUPPORT
SHOWN WITH 7516.172 FTG., 3/8 FLEX. SUPPLIES,
7723.018 DRAIN ASSEMBLY, 1-1/4 x 1-1/4 O.D."P" TRAP

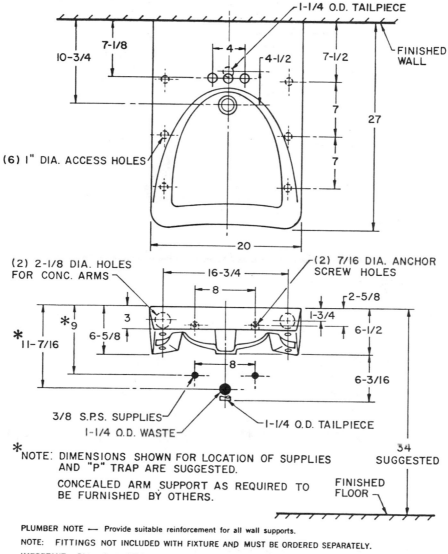

NOTE: DIMENSIONS SHOWN FOR LOCATION OF SUPPLIES
AND "P" TRAP ARE SUGGESTED.

CONCEALED ARM SUPPORT AS REQUIRED TO
BE FURNISHED BY OTHERS.

PLUMBER NOTE — Provide suitable reinforcement for all wall supports.

NOTE: FITTINGS NOT INCLUDED WITH FIXTURE AND MUST BE ORDERED SEPARATELY.

IMPORTANT: Dimensions of fixtures are nominal and may vary within the range of
tolerances established by ANSI Standards A112.19.2. M

These measurements are subject to change or cancellation.
No responsibility is assumed for use of superseded of voided leaflets.

American Standard

Rough-in courtesy of American Standard.

AKRON SERVICE SINK   7696.016

ENAMELED CAST IRON
SHOWN WITH
8344.111 FAUCET   7798.176 TRAP STANDARD

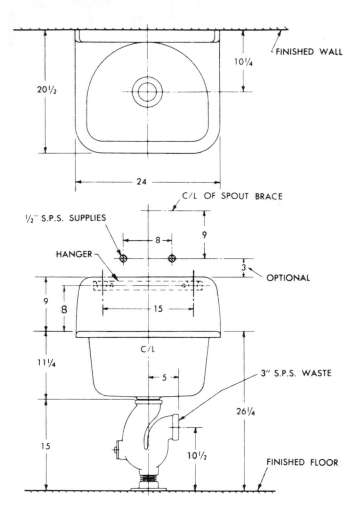

FINISHED WALL

10¼

20½

24

C/L OF SPOUT BRACE

½" S.P.S. SUPPLIES

HANGER

8

9

3

OPTIONAL

9

8

15

C/L

11¼

5

3" S.P.S. WASTE

15

26¼

10½

FINISHED FLOOR

NOTE: Fittings and trap standard not included with fixture and must be ordered separately.

PLUMBER NOTE — Provide suitable reinforcement for all wall supports.
IMPORTANT: Dimensions of fixtures are nominal and may vary within the range of tolerances established by ANSI Standards A112.19.1.M
These measurements are subject to change or cancellation. No responsibility is assumed for use of superseded or voided leaflets.

American Standard

Rough-in courtesy of American Standard.

LEDGEMONT LAUNDRY SINK

7601.016
ON
7602.030

ENAMELED CAST IRON
SHOWN WITH— RELIANT   TOP MOUNT
OR BOTTOM MOUNT FAUCET &
4362.034 DRAIN PLUG

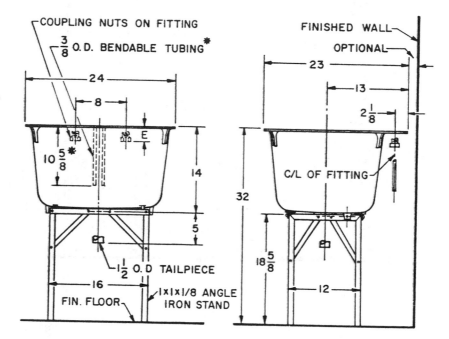

FAUCET		E
RELIANT	4205.041	✳
TOP MOUNT	4150.132	2-1/4
BOTTOM MOUNT	4146.130	4

NOTE: CONNECTIONS TO WALL SUPPLIES MUST BE MADE BELOW TRAY.

NOTE: FITTINGS AND 7602.030 FRAME NOT INCLUDED WITH FIXTURE AND MUST BE ORDERED SEPARATELY.

IMPORTANT: Dimensions of fixtures are nominal and may vary within the range of tolerances established by ANSI Standards A112.19.1. M

These measurements are subject to change or cancellation.
No responsibility is assumed for use of superseded of voided leaflets.

American Standard

Rough-in courtesy of American Standard.

WASHBROOK URINAL   6501.010

VITREOUS CHINA — SHOWN WITH SLOAN 186
OR DELANY 451VB FLUSH VALVE

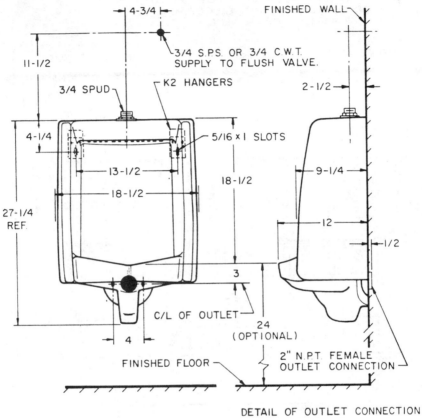

PLUMBER NOTE — Provide suitable reinforcement for all wall supports.

NOTE: FLUSH VALVE NOT INCLUDED WITH FIXTURE AND MUST BE ORDERED SEPARATELY.

IMPORTANT: Dimensions of fixtures are nominal and may vary within the range of
tolerances established by ANSI Standards A112.19.2.M

These measurements are subject to change or cancellation.
No responsibility is assumed for use of superseded of voided leaflets.

American Standard

Rough-in courtesy of American Standard.

ROUGH IN FOR DRINKING FOUNTAIN
MODEL SW13A

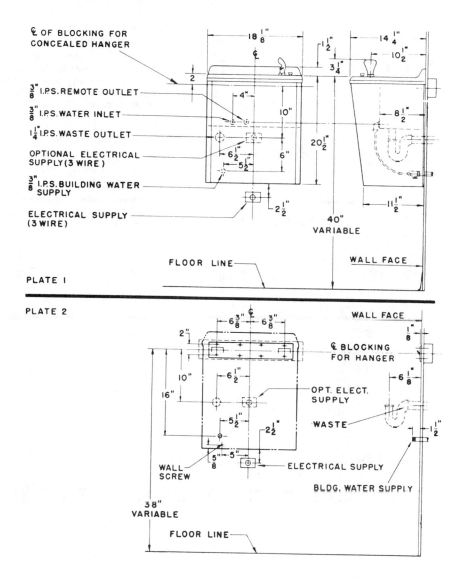

PLATE 1

PLATE 2

NOTE: RECOMMENDED MOUNTING HEIGHTS

Cooler can be mounted at conventional heights to accommodate small children or adults. Recommended height for adults is 40" from floor line to basin rim. (Center line of waste outlet is 28" from floor line.)

Recommended height for elementary school children is 28" from floor line to basin rim. (Center line of waste outlet is 16" from floor line.)

Any change made in mounting height of cooler MUST also be made in height of waste outlet and water inlet from floor line.

THE HALSEY W. TAYLOR CO.   WARREN, OHIO

subsidiary of KING-SEELEY KST THERMOS CO.

CAUTION

For installation when the dishwasher is to be left unused in freezing temperatures: Turn off electrical power. Shut off water supply at the hand valve. Remove lower panel. Disconnect both inlet and outlet lines at the DRAIN valve. This permits water to drain from the dishwasher tank, pump and drain line. Disconnect both the inlet and outlet lines at the FILL valve. Make provisions to control water drained from unit. Reinstall lower panel. Turn on electrical power. Then, push door handle down to lock position and push "Normal Cycle" button. Let dishwasher run through first pre-rinse only (approximately four minutes) to drain all water trapped in the dishwasher. Turn off electricity. To complete installation later:

Connect drain and fill valves—both inlet and outlet.
Reinstall lower panel.
Turn on water and electrical supply. Then push button for cycle desired.

DETAILS AND CONNECTIONS

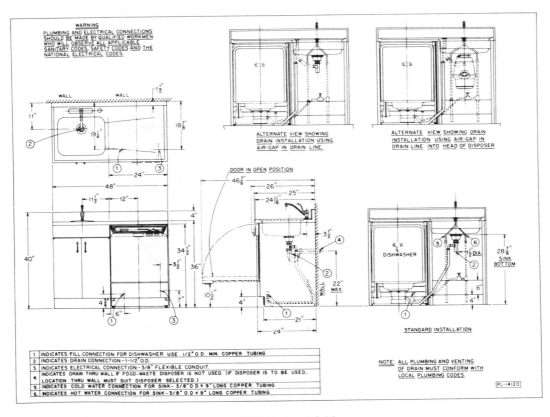

KDC-38 & KDS-38
BASIC DIMENSIONS

KITCHENAID DIVISION **HOBART** TROY, OHIO 45374

GALLERIA WHIRLPOOL
HIGH-GLOSS ACRYLIC
COMPLETE WITH POP-UP C.D. & O.
SHOWN WITH 8900.SER. DECK MOUNTED FITTING

2718.302

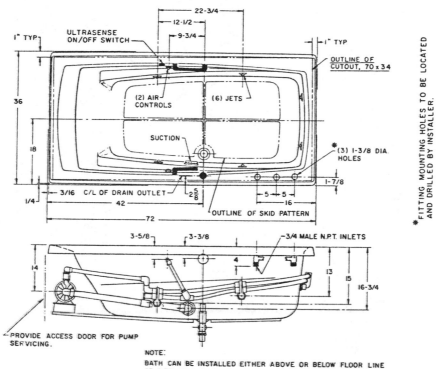

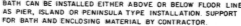

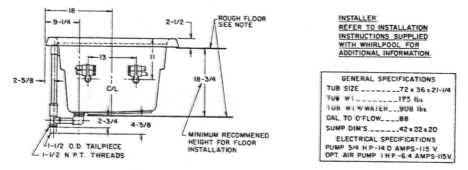

PROVIDE ACCESS DOOR FOR PUMP SERVICING.

NOTE:
BATH CAN BE INSTALLED EITHER ABOVE OR BELOW FLOOR LINE
AS PIER, ISLAND OR PENINSULA TYPE INSTALLATION. SUPPORT
FOR BATH AND ENCLOSING MATERIAL BY CONTRACTOR.

INSTALLER:
REFER TO INSTALLATION
INSTRUCTIONS SUPPLIED
WITH WHIRLPOOL FOR
ADDITIONAL INFORMATION.

GENERAL SPECIFICATIONS
TUB SIZE _____72 x 36 x 21-1/4
TUB WT._____175 lbs.
TUB WT. W/WATER___908 lbs.
GAL. TO O'FLOW.____88
SUMP DIM'S._____42 x 22 x 20
ELECTRICAL SPECIFICATIONS
PUMP 3/4 H.P.-14.0 AMPS-115 V.
OPT. AIR PUMP 1 H.P.-6.4 AMPS-115V.

PLUMBER NOTE — Provide suitable reinforcement for all wall supports.

NOTE: FITTING NOT INCLUDED WITH FIXTURE AND MUST BE ORDERED SEPARATELY.

IMPORTANT: Dimensions of fixtures are nominal and may vary within the range of
tolerances established by ANSI Standards Z124.1.

These measurements are subject to change or cancellation.
No responsibility is assumed for use of superseded of voided leaflets.

American Standard
5

B443

JULY 1988

Rough-in courtesy of American Standard.

EMERGENCY USE PLUMBING FITTINGS
(COLLEGE CHEMISTRY LABORATORY)

The High Flow Shower is installed without a floor drain or encloser.

The dual Eye Wash Spray Heads are installed on a wall-hung lavatory.

Index

The packet attached to the back cover contains 7 *residential* prints and 11 *commercial* prints.

RESIDENTIAL PRINTS

SHEET 1	Front	Plot Plan
	Back	Basement Plan
SHEET 2	Front	Floor Plan
	Back	Northeast Elevation
SHEET 3	Front	Southeast Elevation, Northwest Elevation
	Back	Southwest Elevation
SHEET 4	Front	Kitchen Fireplace Details

COMMERCIAL PRINTS

SHEET 1	Front	Plot Plan and Index
	Back	Basement Floor Plan
SHEET 2	Front	First Floor Plan
	Back	Second Floor Plan
SHEET 3	Front	East Elevation, Store Front Details
	Back	West Elevation, Roof Plan, Schedules
SHEET 4	Front	North Elevation
	Back	South Elevation, Window Details
SHEET 5	Front	Longitudinal Section Framing Details
	Back	Transverse Section Wall Details
SHEET 6	Front	Retaining Wall Details